MAGDALEN
ASCENSIO

"Annine van der Meer has brought together strands of historical evidence with new channeled material. She has done this with all the genius and diligence of her quest for the truth about Mary Magdalene, has brilliantly succeeded in filling the gaps in her story, and has shown us the path to the ascension experience. This book is revelatory, powerful, and healing for the modern psyche lost in the babel of social media and the false ideology of scientific materialism. I regard it as the summit of all her work because it brings together the visible and invisible dimensions, the Above and the Below, in a wonderful synthesis. What an achievement!"

ANNE BARING, PH.D. (HONS), AUTHOR OF
THE DREAM OF THE COSMOS: A QUEST FOR THE SOUL

"My earliest memory of an alternative vision of Christianity's Mary Magdalene-as-prostitute was in the late 1960s when *Jesus Christ Superstar* was performed on Broadway; its radical drama and vibrant music were spellbinding and revolutionary. In the decades since, variations on the theme have been expressed through scores of biographical novels and works of creative nonfiction—joining classic biblical approaches to the subject but positing an actual marriage between Mary Magdalene and Jesus—books like *The Woman with the Alabaster Jar, The DaVinci Code,* and *Holy Blood, Holy Grail.* Annine's book *Magdalene's Ascension* goes a step further by digging seriously into the esotericism of the Essenes, the esoteric Western magical tradition, plus the disciplines of shamanism, women's spirituality, and theosophical philosophy. A work well done!"

VICKI NOBLE, HEALER AND TEACHER, AUTHOR OF
MOTHERPEACE, SHAKTI WOMAN, AND *THE DOUBLE GODDESS*

"Using a multifaceted approach that includes canonical and apocryphal as well as Gnostic texts, historical documents, and alternative sources from channeling and hypnotic regression, Annine van der Meer offers readers a remarkable and convergent visionary synthesis of the spiritual significance of Mary Magdalene through seven complementary portals. Central to the book is the ascension process so vividly described in *The Gospel of the Beloved Companion*, but also the return of the Sacred Feminine, represented in the mystical Gnostic way as complementary to outer forms and structures. There are many rare gems of wise insight throughout this inspiring book that speak deeply to the inner needs of our time."

DAVID LORIMER, GLOBAL AMBASSADOR AND
PROGRAM DIRECTOR OF THE SCIENTIFIC AND MEDICAL NETWORK
AND GALILEO COMMISSION CHAIRMAN

"Of all the books on Mary Magdalene, this is by far the most erudite, comprehensive, and inspiring. After 2,000 years of suppression of this great figure who did so much to influence the life of Jesus and early Christianity, Mary Magdalene is resurfacing in our time. Annine van der Meer has provided the must-read book for all those who wish to understand the profound way in which Mary Magdalene has contributed to the spirituality that lies at the heart of the Western mystical tradition."

JIM GARRISON, FOUNDER AND PRESIDENT OF
UBIQUITY UNIVERSITY AND AUTHOR OF
THE PLUTONIUM CULTURE

MAGDALENE'S ASCENSION

Mary's Journey to Becoming Light

A Sacred Planet Book

ANNINE VAN DER MEER, PH.D.

Bear & Company
Rochester, Vermont

Bear & Company
One Park Street
Rochester, Vermont 05767
www.BearandCompanyBooks.com

Bear & Company is a division of Inner Traditions International

Sacred Planet Books are curated by Richard Grossinger, Inner Traditions editorial board member and cofounder and former publisher of North Atlantic Books. The Sacred Planet collection, published under the umbrella of the Inner Traditions family of imprints, includes works on the themes of consciousness, cosmology, alternative medicine, dreams, climate, permaculture, alchemy, shamanic studies, oracles, astrology, crystals, hyperobjects, locutions, and subtle bodies.

Originally published in 2023 in Dutch by Pansophia Press under the title *Maria Magdalena, Poort naar de Nieuwe Tijd: Wegwijzer naar het Licht*
First U.S. edition published in 2025 by Bear & Company

Cataloging-in-Publication Data for this title is available from the Library of Congress

ISBN 978-1-59143-526-6 (print)
ISBN 978-1-59143-527-3 (ebook)

Printed and bound in India by Nutech Print Services

10 9 8 7 6 5 4 3 2 1

Text design by Priscilla Harris Baker and layout by Debbie Glogover
This book was typeset in Garamond Premier Pro with Corporate A Pro, Gill Sans MT Pro, Grand Central, and Worthington Arcade used as display fonts

To send correspondence to the author of this book, mail a first-class letter to the author c/o Inner Traditions • Bear & Company, One Park Street, Rochester, VT 05767, and we will forward the communication, or contact the author directly at **pansophia-press.nl**.

*To my dear Dutch friend and respected colleague, drs.
Karin Haanappel, for her tireless efforts to bring the divine
feminine back into the world and as a thank-you for the
many years of fruitful cooperation.*

CONTENTS

MARY MAGDALENE AND MY LIFE-CHANGING EXPERIENCE

While writing my latest book, *Mary Magdalene Unveiled,* in 2020, my life gained momentum. I came into contact with channeled sources, which seemed credible to me; they touched me. Mary Magdalene provided a life-changing experience, which is how I experienced it. What was happening? Books came my way that don't fit within the framework of traditional science. They were books that were not based on historical texts and sources but that mentioned memories of people in regression, memories from the time of Mary Magdalene and Jesus. These people in regression were eyewitnesses. This alternative information had a positive effect on me. I had a much clearer insight and overview of what happened around the beginning of the Christian era. I received answers to questions I had been asking myself all my life as a scientific researcher.

The people who remember a life associated with Mary Magdalene and Jesus paint a much more human picture of both of them. Although there are a lot of similarities, there is one big difference from the historical writings from classical antiquity, which disseminate male-oriented and often extremely misogynistic information. These people in regression seem to link the fragmented and sometimes contradictory information from ancient historical sources into a whole. When you read the experiences of these people in regression, the whole picture seems to form. You gain an overall view, a helicopter view. They seem to provide important missing puzzle pieces with which a much larger part of the puzzle can be laid out. I have been working with historical sources, also from the period of classical antiquity, for almost fifty years. It is beautiful but also frustrating work. There are so many contradictions in the minimal material that remains from that time. For example, the biblical texts (of which only about 15 percent survived) were edited and reworked.

As a researcher, I felt I'd suddenly received this wealth of information thrown into my lap, and the obvious questions arose: Are the alternative sources reliable? Can they complete the picture? Can they be added to the historical source material? All this challenged me to explore the existing academic boundaries and new paths that fell outside of the traditional scientific framework. I serve the groundbreaking postmaterialistic new science. I had planned to put the traditional and alternative sources side by side and compare them and have already integrated numerous scientific disciplines in previous books, illuminating a certain subject from many angles. In doing so, I used the interdisciplinary and comparative method, which I will also do in this book and expand upon it further by including alternative sources of regression and channeled material. Decide for yourself whether the alternative sources that I propose below provide the book with added value. For me personally, the following is beyond dispute.

Mary Magdalene has been my guide. I am convinced that she opened new avenues for me. She broke me open. She has led me out of the traditional academic straitjacket and has broadened my view. That she does this has to do with the New Era that we are entering. Her living *gnosis* needs to be transferred again into the twenty-first century. She invites you and me to give wings to our consciousness and to explore the Other World—within us, in the Earth, outside us and outside the Earth—in free flight. On to the New Earth, the New World, and the New Time. Have a good trip!

1

NEW ROADS OPEN UP

On Valentine's Day in 2021, my latest book *Mary Magdalene Unveiled* was published. It is based on *The Gospel of the Beloved Companion: The Complete Gospel of Mary Magdalene*. After extensive study I became convinced that this gospel from the first half of the first century was written by Mary Magdalene. It fits into a Marian tradition to which other early Christian texts also belong, although there you deal with small fragments and not with a consistent whole.[1]

On January 10, 2022, a journalist came to interview me for the spring issues of a magazine.[2] He asked me if I was risking my reputation by writing a book on a dubious gospel. He mentioned that there was no photograph of the document of *The Gospel of the Beloved Companion*, so no hard proof of the original manuscript was available. Its whereabouts in France were unknown, so why couldn't it be a forgery? According to him, I was risking my reputation by taking such a controversial text only known in modern translations seriously.

I answered him as follows.

My intuition, trained by a little less than fifty years of research into ancient pre- and early Christian texts, tells me that this text is pure and true, and this has been confirmed by measurements from spiritual

---

1. A great deal of source material about the female-friendly, early Christian past has been lost, written down, and manipulated; one assumes 85 percent. That makes the Gospel of Miryam unique.
2. Journal OMG or O(ther)M(e)G(od), spring edition 2022.

radiesthesia (see chapter 20).[3] Mary's Gospel has been preserved within a closed Cathar community in western France since the twelfth century. The Cathar Laconneau community has since protected, preserved, and passed it on from generation to generation. Cathars speak the truth; they don't lie.

What a relief it was to work with an unblemished text in 2020 as I wrote my book *Mary Magdalene Unveiled*, in which a woman speaks. She truthfully reports on the public life of Yeshua the Nazorean, herself as coteacher and successor, and their male and female followers. Mary Magdalene in this text calls herself Miryam the Migdalah. Miryam is her original Hebrew name, and Migdalah or Magdala means "the toweress" or "the elevated one"—this is a function, not the name of a town. She describes herself as Miryam of Bethany; she was Yeshua's wife and successor.

At the end of her Gospel, Miryam takes a shamanistic soul journey through the spheres or dimensions as she makes a heavenly journey or ascent. On her way to the light, she transcends the world of duality and meets a lady, the Queen of Heavens, on the eighth level. The Gnostic Christians talk about her realms as the Kingdom of Light, the Pleroma or Fullness. They believe that humans have descended from the light world into the shadow world of 3D as children of the Light. In the gnostic view, Mary Magdalene is the wayshower who shows the way home.

The Regression Reports

In earlier books about the divine feminine and spiritual women's studies, I pioneered paths outside the traditional scientific framework.

---

3. On March 27, 2021, a measurement by DJR-Advies (in English: DJR Advice), a group specialized in spiritual radiesthesia, found *The Gospel of the Beloved Companion* to be authentic. Radiesthesia is a form of dowsing by use of a pendulum to measure Bovis values and, among other things, the amount of truthfulness in and of books. Their email read: "Through a form of spiritual radiesthesia, the authenticity of the Gospel recorded by Mary Magdalene is confirmed. There should be no doubt whatsoever about the authenticity of this Gospel."

Again, in this book, I am undertaking an experiment in which I compare the traditional sources about Mary Magdalene and Yeshua with data from regression reports.

I have taken note of this eyewitness material and, as said in the preface, find it very interesting. But I need to know if I can really trust these reports. Therefore, while using the comparative method, I need to compare the information of antique sources with the regression material. I ask: Do accounts of ancient authors match eyewitness accounts of people in regression? Do the latter add anything? Do we get a more insightful, complete, and human picture of that time? If so, it is desirable no longer to exclude this material. This is my working hypothesis.

The Dutch theologian Joanne Klink (1918–2008) was a controversial pioneer in her time who knocked down many dogmatic theological houses. Even before her controversial book *The Unknown Jesus* in 1996, in which she reported about regression reports,[4] she had already caused a stir in 1990 with her book *Earlier When I Was Big: Far-Reaching Memories of Small Children*.[5] In this she wrote that small children are still open to "the other world." They spoke of inhabiting a world of light before their birth; then they got a push and descended.[6] Klink interviewed a religious educator who worked with small children. She reported that a child said: "On a ray of light I came down from heaven and on the same I go up again. My belly was still transparent."[7]

In one of my previous books, I had already processed data from Klink, which concerned memories of children's past lives. In it I had referred to the regression research of Helen Wambach and Carol Bowman.[8] But what made a big impression on me was the monumental work of the scientific founder and pioneer in this field, the psychiatrist

---

4. Klink, *De Onbekende Jezus* (*The Unknown Jesus*), epilogue, 237.
5. Klink, *Vroeger toen ik groot was*, 95.
6. Klink, *Vroeger toen ik groot was*, 54.
7. Klink, *Vroeger toen ik groot was*, 108.
8. Helen Wambach, *Life Before Life: Is There Life Before Birth?*, Toronto: Bantam Books, 1979; Carol Bowman, *Return from Heaven: Beloved Relatives Reincarnated within Your Family*, New York: HarperTorch, 2003.

Ian Stevenson (1918–2007). He was the author of approximately three hundred articles and fourteen books about reincarnation.[9] In his 1997 book he reported two hundred cases where moles and birth defects seemed to correspond in some way to a wound from a past life that the child remembered.[10]

In her book *The Unknown Jesus* Joanne Klink included lengthy quotes from regression protocols from people who had known Jesus in an Essene past life. They were regressed by Dolores Cannon, an American. I was flabbergasted when I read this. A passage from chapter 4 ("An Eyewitness Report") made an even deeper impression on me:

> A few decades after the Dead Sea Scrolls were discovered, something different was discovered in our Western culture and mindset. Not in caves or under the sands of the desert or in the dusty attics and cellars of ancient monasteries, but in the human being itself. This was due to a particular technique of psychotherapy, in which it turned out to be possible to bring people into such a deep relaxation that they were able to go beyond the here and now, yes even outside the time of their lives. Unexpectedly personal memories emerged that went back much further than birth or even conception. They were more than memories, but real re-experiences of incidents and situations from the sometimes distant past that turned out to be related to traumas or relationship problems in this life. Awareness appears to lead to healing. This regression therapy has been a common, albeit still alternative, therapy for years. With slightly slower brain waves than our daytime consciousness, the alpha waves, these regressions generally take place, while still being in the here and now at the same time. If one goes deeper in relaxation to the theta waves,

---

9. Ian Stevenson, *Twenty Cases Suggestive of Reincarnation*, Charlottesville, VA: University of Virginia Press, 1974; *Cases of the Reincarnation Type*, four volumes, Charlottesville, VA: University of Virginia Press, 1975–1983; and *European Cases of the Reincarnation Type*, Jefferson, NC: McFarland & Co., 2003.
10. Ian Stevenson, *Reincarnation and Biology: A Contribution to the Etiology of Birthmarks and Birth Defects*, Vols. 1 and 2, Westport, CN: Praeger, 1997.

then in his experience he is no longer here and now, but completely in the past, ignorant of things and situations in our own time. In general, this regression method is used when it is necessary to better understand or let go of soul problems in life. But occasionally it happens that the memories are historically interesting and that people spend a long time on them, sometimes months, to tell about one particular life. . . . Skeptics are only too happy to argue that this will of course be based on fantasy. Yet they do not bother to take note of these protocols. It turns out that the story is told with all kinds of historical details, more elaborate than a history book, in which there is not a single sentence that does not fit into the historical pattern of that period. That is all the more striking when neither therapist nor client has ever heard anything about that time.[11]

Thus it happened that Cannon's client, after memories of many lifetimes, said that her name was Suddi Benzamare, and she had been a Torah teacher in Qumran. Neither the therapist nor the client had ever read anything about Qumran or the Dead Sea Scrolls, and even the word *Torah* was unfamiliar. Cannon chose not to read anything more about it for three months while her client spoke about her experiences, which provided an eyewitness account of the life and thinking of the Essenes in Qumran at the beginning of the Common Era.

Klink wrote in 1996 that "it would be important if scientists who stare at the texts of the caves at Qumran to decipher them, would prepare themselves to read the text of these regressions, which could solve many of the riddles we now face. And on closer inspection, they might notice that virtually all of the Essene features described by first-century historians are reflected in this regression."[12]

I could have accepted the latter at the time and put it in my pocket, but these things were completely new to me. I thought, "How fascinating and courageous that Klink mentions this," but I also thought, "This

---

11. Klink, *De Onbekende Jezus*, 31–32.
12. Klink, *De Onbekende Jezus*, 30–31.

is going a bridge too far for me." And I left the regression protocols for what they were.

Thus I ignored channeled messages because I felt they were too colored by the receiver's filter of beliefs. I hadn't wanted to use it in research because I knew that all information has to go through the channel's filter, which could potentially distort it.

But now I know that the vibration of the Earth is rising and that as a result the information from other dimensions has become more accessible to people on Earth, so I've taken a different attitude toward regression reports as well as some channeled messages that feel pure to me.

I cite Klink again because she summarizes clearly and also in short staccato sentences, as if in a hurry, what information the three months of regression back to that Essene life of Suddi Benzamare yielded.

The white robes, that they [the Essenes] were in a sense a secret order and were not allowed to tell certain things to strangers. That they were against animal sacrifice, had no slaves, had disciplinary measures, various degrees of initiation, gave everything to the community and had no money, that they took daily baths for purification, bathed at sunrise, had a solar calendar and not the lunar calendar of the Jews. That there was a children's class, where children were educated, that there were also women, but not all were married, that they practiced astrology, calculated the birth of the Messiah and prepared his way on Earth, believed that death was a transition and that the soul is immortal, had faith in providence, could foresee the future, paid much attention to prophecies, that angels were in their teaching, and a special focus on the prophet Enoch. The whole report is marked by a vividness of being an eyewitness. This is evident in the small details. The 22-year-old woman who was a teacher in one of her previous lives, Suddi Benzamare, scratches her cheek during the memory and exclaims that the mosquitoes sting again. Or suddenly starts talking in such an accent that it was almost unintelligible. As a child spoke with a child's voice, but when he grew old, with a tired voice. Old, that was 45 years. When he talks about reading the scrolls, the hand points from right to left. At one point, the therapist gave a ballpoint pen. Although the woman was left-handed, she took it with the right hand and looked at the thing, not knowing what it was. To write, she was told. So then she/he started writing from right to left.[13]

It's time to turn our attention to therapist Dolores Cannon and the methods she used to regress clients.

Dolores Cannon and Her Regression Method

American therapist Dolores Cannon (1931–2014) explains her method in the beginning of her book *Jesus and the Essenes*.[14] She calls herself a "regressionist," which is a modern term for someone who specializes in past life regressions.[15] She started developing her own method in 1979. In a regression it is crucial not to suggest anything but to let the process proceed spontaneously. Just as a hypnotist would, she does not tell a person being regressed to go in a particular direction. The initiative lies with the person in the process itself.

Cannon argues that every person goes into a trance state at least twice a day. In the evening before falling asleep and in the morning before fully waking up. During the day this can also happen when driving a monotonous car, or listening to a boring sermon or lecture, and at countless other times. It is therefore easy for every human being to switch more or less unconsciously to changing states of consciousness.[16]

Many people were willing to be guinea pigs when Cannon developed her method. In fact, 95 percent of people all remembered a past life whether they were uneducated, highly educated, or skeptical of reincarnation. People often don't believe in it, but it happens to them anyway.[17]

---

13. Klink, *De Onbekende Jezus*, 31–32. See also Cannon, *Jesus and the Essenes*, 47.
14. Cannon, *Jesus and the Essenes*, 6–8.
15. Cannon, *Jesus and the Essenes*, 6.
16. Cannon, *Jesus and the Essenes*, 6.
17. Cannon, *Jesus and the Essenes*, 7.

Cannon was not interested in statistics, but in people, their stories, and their problems.

Cannon explained in the beginning of *Jesus and the Essenes* that there are different stages of hypnotic trance. They are registered in laboratories with different measuring instruments. How deeply the person is in trance can be deduced from the physical reactions and the answer the therapist gets to the question posed. The deeper the trance, the more the person can remember. Even in a light trance state people can remember a past life. When this happens, people think nothing unusual is happening; they think they are fully conscious and cannot understand where the information is coming from. Because the conscious mind is still very active, they think they're imagining it or that it's their own fantasy playing up. In the lighter trance forms it is as if when remembering a past life they are watching a movie. When the hypnosis deepens, the person participates in the film of the past life. When they see everything through the eyes of the person they once were and also experience emotional reactions, they go even deeper. When the conscious mind becomes less active, they become more involved in what they see and experience.[18]

Cannon considered the sleepwalker's condition to be a form of deep trance. The person becomes the old personality and fully revives that life. In addition, they possess no memory of any other time. In every way they become that person who lived hundreds or thousands of years before. They can only tell what they know. When it comes to peasants, they don't know what happens in the king's palace and vice versa. They are often unaware of the events reported in any history book when those events did not affect their personal lives at the time. When they wake up, they remember almost nothing. They think they have fallen asleep; every scene left in the conscious mind resembles the fading fragments of dreams.

In a sleepwalker state, they can give a lot of information because they are completely that personality who lived in that particular period. It's fascinating to watch someone in such a deep trance change and

---

18. Cannon, *Jesus and the Essenes*, 7–8.

assume the manners and voice of someone else completely. This type of deep trance is not easily achieved by most people. Cannon determined experimentally that in her practice, only one in twenty people could achieve this state. Most people don't dare to indulge in this deep state of trance; they hold up the safe walls of self-defense. Cannon found that trust must grow before the walls will collapse. The person must realize that he or she is perfectly safe. Cannon's technique relies on her ability to help a person build trust and cooperation within one's mind and not take control of the process.[19]

Most of the lives remembered are not spectacular. Cannon never found a Cleopatra or a Napoleon. She considered it a sign of truth that most people remember lives that are ordinary and routine. There are no fantastic stories. Sometimes there is an exciting life in between, but the dull lives far outnumber them. It's about soldiers who never went to war, Native Americans who lived peaceful lives instead of fighting the white man, or farmers and early settlers who knew nothing but grueling work, worry, and unhappiness. What influenced most regressions was not the actions or adventures of people, but the human emotions they experienced. When someone wakes up from a trance with fresh tears still flowing down their cheeks after remembering an event that took place more than two hundred years ago, no one can tell them that it is a fantasy. It's like recalling childhood memories. One can properly place the event and see how it has affected this present life and learn from the recalled memory.[20]

At a gathering one evening, Dolores Cannon met Katherine (Katie to her friends) Harris, then twenty-two years old. This name is a pseudonym, partly because of the Christian background of Katie's parents. Katie had registered with Dolores for a voluntary regression therapy to past lives. Born in 1960, she had roamed through the United States with her parents. After two years of high school, her family moved back to Texas when she was sixteen. She could no longer cope with this

---

19. Cannon, *Jesus and the Essenes*, 8–9.
20. Cannon, *Jesus and the Essenes*, 9–10.

umpteenth adjustment to a new school system and she dropped out of school. The highly detailed geographic knowledge she displayed under hypnosis could not have been acquired in school. Her knowledge did not come from books. At seventeen she joined the Air Force and completed a two-year computer training course. At that time she had never been outside the USA, but while in deep trance she described many strange places in minute detail. When she was nineteen, Katie and her family moved to a Midwestern town and met Cannon. In the meantime, she worked in an office using her computer skills. She never went to a library to get background on the information that came out during the regressions.[21]

Katie was not a normal trance person. She quickly fell into a deep trance, and she tasted, smelled, and experienced emotions, but she did not remember anything on awakening. In Cannon's eyes, she was the perfect person to fall into the deep sleepwalker trance. After the first two sessions, Dolores instructed her to go into a deep trance after saying a password. She was not told to go to any place or time period. The information came spontaneously. After a month, Dolores decided to work more systematically and to arrange the regressions in chronological order. She began to go back in time by leaps of a hundred years to discover how many lives Katie had lived. During these sessions they also ended up in a region where the soul goes after death—an intermediate region between death and life. After going back a lifetime each week, they unraveled twenty-six separate lives until they came to the time at the beginning of the Christian era. It was about a life as a woman, then as a man, then rich, then poor, then intelligent, and then uneducated. Each life in a deep trance gave Katie a wealth of detail about the religious dogmas and cultural customs of the time. According to Cannon, even a scholar trained in history and anthropology could not have given the incredible detail she gave, so her conclusion was that this knowledge came from elsewhere. They must come from the large "computer files" in her unconscious mind.[22]

---

21. Cannon, *Jesus and the Essenes*, 12–14.
22. Cannon, *Jesus and the Essenes*, 15.

Regressing to the year 400 CE they met a doctor from Alexandria; at around 300 a monk from Tibet. Around 200 CE they came to the life of a deaf-mute girl with whom it was impossible to communicate. Then they came to the Christian period at which point Dolores stopped going further back in time. Instead, she kept coming back to the life of Suddi, one of Jesus's teachers. The person called Suddi had a strong accent. Cannon said his voice changed over the years; it was young and strong as a child, gradually maturing until he spoke in the extremely tired voice of an old man. Information also came from his soul after his death.

Suddi said his name was Benzamare right from the first meeting. When Dolores asked what she should call him, he gave permission to call him by his first name, Suddi. He said he lived in the hills near a community two days' walk away. When Dolores asked him for the name of that community—a normal question according to Dolores—Suddi changed. There was not the usual openness as in previous sessions. He became suspicious and said, "Why do you want to know?" After much hesitation, he finally said that it was called Qumran. At that moment, Dolores said nothing; only later would she look into this. When asked about his profession, Suddi said, "I study the books of the Torah and I study the law, the Hebrew law." In the second part of Cannon's book *Jesus and the Essenes,* it appears that Suddi spent four to five years with two special students. After many detours, Dolores found out their names, which Suddi at first did not want to reveal: Benzacharias and Benjosef, or John and Jesus. He instructed them in the Hebrew law from the ages of eight to twelve and a half.[23]

After that, Cannon came back to Suddi's life for thirteen sessions over a period of more than three months after having worked with Katie for nine months. If Katie had regressed to Suddi's life first, Cannon would have considered it fantasy. But after nine months of contact, Cannon knew the power of Katie's memory. A strong bond of trust had

---

23. Cannon, *Jesus and the Essenes,* 220. See chap. 20, "Jesus and John: Two Students at Qumran," 217–221.

been developed, and the possibility that Katie would lie or fantasize was excluded by Cannon.

After the first few sessions, Katie would listen the tape-recorded session. But upon awakening after later sessions, she would instead simply ask "Where did we go today?" As the material about Jesus came out, it seemed to be bother her, all the more so when things came up that were controversial and contradictory to the Bible. This had to do with Katie's parents having a Christian background; they were members of the Assemblies of God, and Katie did not want a break or separation from her parents. After thirteen sessions and over a year of working with Cannon, Katie decided not to have any further sessions.[24]

Cannon decided that the material was so important that she had to publish it in book form. Her book consisted of two parts and had two objectives: First, to provide information about the Essenes, about whom little was known at that time. Suddi provided information about their habits and living conditions, and the information about the construction of Qumran was so exact that the descriptions corresponded to archaeological excavations.[25] Second, Suddi gave information about the life of Jesus. It was clear that Jesus was associated with the Essenes—he was described from the eyes of a loving teacher, Suddi.[26]

The British Group:
Stuart Wilson and Joanna Prentis

Dolores Cannon's books became bestsellers, including in England, and Cannon trained British therapists in the method she developed. As mentioned, during and after the writing of *Mary Magdalene Unveiled*, people crossed my path who brought books to my attention about regression accounts. After some hesitation, I ordered one of those books from the British group's series: *The Essenes* by Stuart Wilson and Joanna Prentis.

---

24. Cannon, *Jesus and the Essenes*, 13, 17.
25. Cannon, *Jesus and the Essenes*, 125.
26. Cannon, *Jesus and the Essenes*, 28.

I read *The Essenes* with increasing surprise. I was amazed—just as Joanne Klink had been before me—at the similarities between the information well known to me from ancient authors and the information from these regression reports. My intuition told me that these records are pure and reliable, exactly as I had that feeling and inner knowing when I studied *The Gospel of the Beloved Companion* intensively. Then I had the information measured through spiritual radiesthesia. The measurements showed that the information was "completely pure (see chapter 20)."[27]

Stuart Wilson was known in England for his bestselling baby name dictionary.[28] In the 1990s this British author moved to the Starlight Center in Devon, southern England, founded in 1988, to further develop the healing and transformation of consciousness together with founders Joanna Prentis and her daughter, Tatanya. In the foreword to *The Essenes*, Wilson wrote that they both worked together in the center on past life regressions. Prentis, a qualified hypno- and regression therapist, led the sessions. Wilson elaborated upon the material in the tapes and processed the material into book form. For the book *The Essenes,* seven people went into regression, five of whom had ties to the Essenes and two who were disciples of Jesus with no affiliation with the Essenes. The first four people knew each other during their Essene lives. Three lived in communities west of the Dead Sea; one of them was a trader and traveled constantly. That they knew each other was a key to getting through the Essene obligation of secrecy—a veil of secrecy that the Essenes wove around their communities and activities.[29]

After *Jesus and the Essenes* was printed in England in 1992, Dolores

---

27. As previously stated, a spiritual radiesthesia measurement by DJR-Advies had found *The Gospel of the Beloved Companion* to be authentic. Futher, DJR-Advies conducted a spiritual radiesthesia measurement on January 6, 2022, on *The Essenes,* concluding: "This information is completely pure." Measurements of other books in this series gave the following results: For *Power of the Magdalene,* "This information contains minor deceptions" and for *The Magdalene Version,* "This information is mostly correct."

28. Stuart Wilson, *Simply the Best Baby Name Book: The Most Complete Guide to Traditional and New Names,* London: Pan, 2001.

29. Wilson and Prentis, *The Essenes,* i–iii.

Cannon began lecturing about this material, and for several years she spoke at The Essene Network Summer School in Dorset, England.[30] When she visited the Starlight Center in Devon in the fall of 1999, Wilson and Prentis gave her a copy of the original manuscript. After reading it, she said the work added information about the Essenes to the information she had previously uncovered. She encouraged them to move on. When they met her again in 2002, she encouraged them to expand the manuscript. This led to the publication of their book in 2005 and for this Stuart Wilson expressed his gratitude to Dolores Cannon.[31]

When studying both books about the Essenes, I saw many similarities, such as the obligation of confidentiality with the Essenes—the result of which is they are not allowed to talk freely. Wilson wrote that their book also provided insights into the location of the Essene communities and the existence of a core group. He said it was information that hadn't been revealed so far and that it increased our knowledge of how the Essenes functioned; for example, the balance between the lay brothers and sisters and the priests. Finally, he expressed the hope that the book will encourage others to discover and publish more new regression material.[32]

In her foreword, Prentis introduces herself as a healer, coach, and past-life therapist. Regression therapy can help people overcome fears, phobias, and health problems. All memories of past lives are stored in the emotional body and in the cellular memory. According to Prentis, when these memories are cleared, energy blockages can be resolved and the client can go through life lighter and purer because life is a learning process. We keep repeating the same patterns until we understand the lesson. She found the work fascinating and a good way to learn about history firsthand, although it didn't have to correspond directly to what was in the history books because so much had become disfigured over time.

Like Dolores Cannon, Prentis found that most of her clients have

---

30. Cannon, *Jesus and the Essenes*, 280.
31. Wilson and Prentis, *The Essenes*, ii–iii.
32. Wilson and Prentis, *The Essenes*, ii.

PLEASE SEND US THIS CARD TO RECEIVE OUR LATEST CATALOG FREE OF CHARGE.

Book in which this card was found_____

❑ Check here to receive our catalog via e-mail.

Name_____

Address_____

City_____ State_____ Zip_____ Country_____

E-mail address_____

| Company_____ |
| ❑ Send me wholesale information |

Phone_____

Please check area(s) of interest to receive related announcements via e-mail:

❑ All Books

❑ Ancient Mysteries and the Occult

❑ Psychedelics and Entheogens

❑ New Age Spirituality, Personal Growth, and Shamanism

❑ Holistic Health, Natural Remedies, and Yoga

❑ Sacred Sexuality and Tantra

Please send a catalog to my friend:

Name_____ Company_____

Address_____ Phone_____

City_____ State_____ Zip_____ Country_____

Order at 1-800-246-8648 • Fax (802) 767-3726

E-mail: customerservice@InnerTraditions.com • Web site: www.InnerTraditions.com

INNER TRADITIONS
BEAR & COMPANY

Inner Traditions • Bear&Company

P.O. Box 388

Rochester, VT 05767-0388

U.S.A.

had fairly ordinary lives, but these can be important to them personally. Also, it was rare that the person went into a deep hypnotic state and remembered nothing on awakening. A balanced state of consciousness is much more common. It was as if the normal personality steps aside and they temporarily become the person they were. They also behave like that person and also talk like that person. They are usually semi-conscious of what is happening and rarely remember exactly what they said, which is why Joanna turns on the tape recorder.[33]

During her work as a regression therapist Prentis discovered that some past lives are only important to the client, but others are larger, of common interest, and connect the experiences of others. The latter was the case when she and Wilson explored and investigated an Essene life. A wealth of information was released, and she could never have dreamed of helping to unlock it. It was information that was relevant for the present and the future. Some groups of people incarnate life after life together, and it was a great joy for Prentis to see these old friends meet in regression and recognize each other. In this way a group arose in Devon who relived their life together with Jesus.

Prentis had worked with Wilson for over four years, and they had explored several of his past lives together. In his sessions Wilson always had a detailed memory. According to Prentis, he was one of the best trance people she had worked with. He was able to speak in regression without encouragement for some time, which made the process much easier for her. Over time, Wilson appeared to return to a life in which he was the Essene, Daniel Benezra (son of Ezra). They worked together intensively for six weeks to gain a picture of the main line in that life and would come back to it later in four stages for more details. Beforehand, Wilson remembered (like Prentis herself) an Essene life in the Damascus community of the Essenes, around 100 BCE. The rules and instructions written then are reflected in the Damascus document, which was found in Qumran as part of the Dead Sea scrolls.[34]

---

33. Wilson and Prentis, *The Essenes*, iv, v.
34. Wilson and Prentis, *The Essenes*, 4.

Prentis brought him back to an altered state of consciousness and asked him to connect with the life that immediately followed the life in Damascus. She asked for his name. Wilson replied, "My name is Daniel. That name means you are judged only by God, and God knows what is in your heart. . . . what it means to be judged with a full knowledge of your being . . . what is in your heart and not what appears to be on the outside."

Prentis asked him, "What is your father's name?"

Daniel answered, "My father's name is Ezra, and therefore I am Daniel Benezra. My father was an elder or leader of our community."

Prentis asked him, "What does your father's name mean?"

He replied, "My father's name means 'salvation.' We see salvation as a matter that continues. We are always protected from those things that lead us away from the truth, away from the light. That's why we see it as a step-by-step process, day by day, and we are always saved."

In this answer, Prentis noticed that when Daniel answered, he had used the collective "we" pronoun. She decided to ask him if he belonged to an Essene community. There was a long pause. Finally, he answered, "Yes."

Prentis understood from the long pause that this was a sensitive subject. When she asked him how many people lived in his community, he replied rather sharply, "I'm not allowed to talk about this." Since Daniel remained reserved and closed off during that first session, Prentis assumed that this life would not yield much information.[35]

A week after Stuart's first session about his life as Daniel Benezra, Prentis had an appointment with a female client whose current name at her own request was not to be revealed. She differed from Wilson, who never showed emotions but often had detailed memories. This woman said, "It seems like I feel everything. My memory is a mixture of seeing, hearing, and feeling."

Prentis asked her to go back to her most important life. Once in an altered state of consciousness, Prentis asked her about her name in that lifetime. She replied, "My name is Joseph."

---

35. Wilson and Prentis, *The Essenes*, 4.

Prentis asked, "What is your father's name?" and the answer was: "You don't need to know that." So Joseph was also on his guard. Joanna then asked, "Do people call you Joseph?" The answer was now followed by a pause.

"I am called Joseph of Arimathea."

Joanna wrote that at that moment she almost fell off her chair. After more questions, Joanna suspected that there was also a connection with Jesus and the Essenes. She asked, "Do you know Daniel Benezra?"

Joseph said, "Yes, he is my friend."

Now Joanna had to suppress her cries of joy because she had come into contact with two people who know each other from the time of Jesus; Joseph is even known from the Bible. She decided to bring the two into contact with each other in regression and have them enter into a dialogue with each other. Perhaps there was a way to lift their obligation of confidentiality because apparently the Essenes wanted their story to be told again after two thousand years. She felt that this couldn't be a coincidence but must be a gift from the Holy Spirit.[36]

Prentis noted in her introduction that Wilson did not read Dolores Cannon's book *Jesus and the Essenes* until most of the regression process was over. This was to ensure that he would not be influenced by inside information. To her, the extraordinary aspect was that much of Daniel's story was corroborated by the four others who regressed to remember an Essene's life, including Joseph of Arimathea, Luke, the Silent One, and the one whose name is unknown. Of course there were differences, but they were caused by a difference in perspective.[37]

Chapter 25 of *The Essenes* features Mary Magdalene, and chapter 26 features Jesus and Mary Magdalene.[38] Daniel talked about three groups of Essenes in Palestine: the urban Essenes functioned in the outer circle; they lived with family in the cities and were most adapted to the Jewish norms and values of the time. The inner circle contained

---

36. Wilson and Prentis, *The Essenes*, 8–11.
37. Wilson and Prentis, *The Essenes*, i, vi.
38. Wilson and Prentis, *The Essenes*, 179–183, 184–188.

closed communities who retreated to remote and difficult areas. This information can also be found in the writings of some ancient authors. But Daniel gave absolutely unique information: he distinguished a third core group of which Mary Magdalene was a part. The existence of this group was hardly known to anyone even within Essene circles, restricting the information to only a few leaders of each community.[39]

Daniel categorically maintained that the main reason why the Essene movement founded around 150 BCE was later transplanted from Egypt to Palestine was to prepare for the coming of the Great Teacher or Teacher of Righteousness.[40] It was the task of the core group to prepare this mission in the utmost secrecy and in the smallest detail, hence the veil of secrecy that surrounded the Essenes. They felt threatened because they were a small minority and studied scrolls that are not acceptable in Judaism. Their remote settlements had underground tunnel systems through which they escaped in times of danger. They also developed a defense system around their establishments that would be associated with sound.[41] People who had nothing to do and yet came too close, felt unwell, and tripped. In turn, the Essenes felt little sympathy for the Jews outside the brotherhood. Daniel said, "They chose a wrong path; we see them as sons of darkness, while the Essenes are the sons of light."[42]

The regression reports of the group of seven clearly showed that Jesus and Mary Magdalene often spent time together and worked together. They had a clear spiritual and intimate bond and engaged in deep conversations that others envied.[43] Of the inner group of disciples she was the guardian of the inner mysteries of the New Way;

---

39. Wilson and Prentis, *The Essenes*, 124.

40. Wilson and Prentis, *The Essenes*, 3, 29, 31, 44, 127, 161. This is related to the teaching of Melchizedek, 319.

41. Wilson and Prentis, *The Essenes*, 92; Cannon, *Jesus and the Essenes*, 57; Heartsong, *Anna, the Voice of the Magdalenes*, 151: In France, a kind of magic is used by the Hebrew community on Mount Bugarach to go unnoticed; what this magic entails is not made clear.

42. Wilson and Prentis, *The Essenes*, 12.

43. Wilson and Prentis, *The Essenes*, 180.

she knew the highest wisdom.[44] We find in *The Gospel of the Beloved Companion* the same kind of information: that during the Last Supper, Jesus appointed Mary Magdalene as leader, and that she fulfilled this function after his disappearance from public life.[45]

Joanna Prentis concludes her introduction as follows: In her view, the Essenes were rooted in two-thousand-year-old ancient history, but they were also surprisingly modern in some ways. They knew the dynamics of the higher self and explored multidimensionality. According to Prentis, today everyone can come into contact with their own higher self, to recognize the divine self within, and to become a multidimensional being. She meets more and more people who are making the leap from their heavy, earthly density. That leap gives access to a wider knowledge and to all the gifts and abilities these people had developed in past lives. That is why she heard more and more about wonderful healings. It became clear to her that the absolute core of Jesus and Mary Magdalene's teachings was unconditional love and that we must return to it. Jesus could do extraordinary things, but he also said that we would do even greater things. Prentis concluded that among those they worked with, greater connections of friendship and love developed, as if they had been ignited in this project by the state of grace and unconditional love that Jesus (and Mary Magdalene) radiated. Finally, she asked us to read their book about the Essenes with an open mind and to take in what is acceptable to the reader in their own heart.[46]

After their successful book, *The Essenes*, the collaboration between Wilson and Prentis resulted in several more books about Mary Magdalene. They were eager to learn more about a slow meditative movement practiced by the Essenes and were advised by their guides

---

44. Wilson and Prentis, *The Essenes*, 180.
45. de Quillan, *The Gospel of the Beloved Companion: The Complete Gospel of Mary Magdalene* [35:16–17], 65–66 (hereafter abbreviated *GBC*); *Mary Magdalene Unveiled* (hereafter abbreviated *MMU*), 385–86.
46. Wilson and Prentis, *The Essenes*, vi, vii.; Joanna and Stuart talk about Jesus and Mary Magdalene; they do not use the Hebrew names Yeshua and Miryam.

to call in extra help. While in trance Wilson said he knew an angelic guide who might be able to provide more information. After a long pause, Prentis felt the energy shift and then communication began: "This is Alariel, speaking on behalf of a group of twelve angels, working together from the Melchizedek order. We understand that you have a question about the Essenes, a brotherhood known to us."[47] Then, a detailed answer came that said these movements only applied to the northern group of Essenes, especially those of the people on Mount Carmel, because there was little room to move. Alariel proved himself to be a great support, because whenever uncertainties arose about events in the past, he jumped in and gave a helicopter view.

From the realm of past-life regression we now come to the subject of channeling. Thanks to modern science, I determined that certain regression and channeled reports could be deemed reliable by adding more tools to my scientific toolbox: the study of quantum and interdimensional physics.

An Excursion into Quantum Physics

In this book, I make use of new scientific insights about ascension from quantum physics. Since the 1980s I have owned a blue ring binder entitled "The Science of Ascension," by David Ash. I took that map with me for many moves, though I never got around to reading the map thoroughly because it contained a lot of physics—until I was confronted with Miryam's ascension, that is. I went to read David Ash's binder and ordered his latest book *Awaken*.

British physicist David Ash developed what he called the "physics of the ascension." He wrote in his book *Awaken* that at the age of four he wanted to prove the existence of God through science. When David was six years old, his father, a doctor who also healed people through

---

47. Wilson and Prentis, *Power of the Magdalene*, 14; Wilson and Prentis, *Atlantis and the New Consciousness*, 31–32; Wilson and Prentis, *The Essenes*, 269–295 (see about the Melchizedek order).

the laying on of hands and the power of thought, introduced him to atomic physics. When he was sixteen years old David discovered an old book about the philosophy of yoga from 1905 while exploring the attic of a private library filled with antiquarian books.[48] In it he read that subatomic particles are a kind of vortex of light and not solid matter. He then understood how important the philosophy and knowledge of yoga was. He had already realized that the yogis of old had an insight into the atom as "an insight that can save the sciences from the dead end into which materialism leads them." He decided to dedicate his life to this topic. He argued with countless other natural scientists that scientific materialism is no longer consistent with new evidence from the quantum field. It can be buried.

The modern philosopher and computer scientist Bernardo Kastrup, who holds doctorates in both philosophy and computer science, holds a similar view: materialism is obsolete. Kastrup developed abstract idealism. Together with Bernardo and a few other speakers, I gave a keynote lecture on June 17, 2022, and studied his work beforehand.[49] During his presentation, he showed the dashboard of a blinded cockpit, where many measuring devices record what happens outside the blinded cockpit. Kastrup compared this to life on Earth. Humanity sits in the blinded cockpit and only sees the inside gauges expand during certain activities in the outside world. That doesn't mean you can deny it like the materialists do.

After fifty years of study and research, Ash's final conclusion was that ascension is the ability to accelerate subatomic particles to a higher frequency and dimension.[50] It is about transforming the gross material body made up of carbon into a shining light body, a so-called diamond body.[51] In alchemy you would say it is about turning lead into gold.

---

48. Ash, *Awaken*, 9, 236, 281, 284, 327, for reference to Ramacharaka Yogi (William Atkinson), 379.
49. Bernardo Kastrup, personal communication during a lecture June 17, 2022, on a conference of the Theosophical Lodge in The Hague. See also *Science Ideated*.
50. Ash, *Awaken*, 9.
51. Ash, *Awaken*, 64.

In early 2022 I read an article by Richard Grossinger—an American anthropologist and literary author who has written more than forty books about modern physics and quantum physics—through the Scientific and Medical Network, which brings together groundbreaking scientists from a variety of disciplines. In one of his books, *Bottoming Out the Universe,* he developed his vision of a universe based on consciousness rather than pure matter. He rejected philosophical materialism, which holds that matter is primary in the universe and the only real reality. Modern physicists increasingly assume that there is consciousness behind and in matter. There is no solid bottom in the subatomic particles of the material universe; the tiniest particles prove that the solid material bottom just isn't there. Small quarks continue to shapeshift: They change from one shape to another. They move in waves and fields. Grossinger noted that concepts such as mind and consciousness are lacking in contemporary physics. Consciousness is considered a side effect of brain chemistry. In his search for consciousness and personal identity, Grossinger also included nonlocal and transpersonal forms of consciousness. In his research into the connection between physics and consciousness, he had gathered evidence from past-life regressions and channeling from higher realms.[52]

In his investigation, he stumbled upon the Seth material. Initially, he described Seth as an entity channeled by Jane Roberts from upstate New York. On September 9, 1963, Jane Roberts was sitting quietly at her table when she came into unexpected contact with an incredible amount of information. She said of her experience: "I felt as if knowledge was implanted into the cells of my body so that I could not forget it."[53] Grossinger noted that Jane Roberts held more than 1,500 trance sessions during a period of twenty years until her death in 1994. Many of them were public, during which she spoke on behalf of Seth, who had very specific ideas about consciousness and saw it as all-encompassing.

---

52. Grossinger, "An Unbottomable Void in Mind," 16–18.
53. Roberts, *The Seth Material,* 11–12; Hanegraaff, *New Age Religion and Western Culture,* 28–29n15.

He did not speak of any personal consciousness but of an intrinsic consciousness that precedes matter and gives rise to universes.[54] Grossinger noted that Seth was emerging as an interdimensional philosopher but wondered what to call him. Was he an entity? Grossinger found that to be a misplaced designation. Seth came across more as a transpersonal intelligence or a union of a group of nonphysical teachers pooling their wisdom as an emanation of an awesome consciousness. Seth believed that materialism was just one thoughtform among others.[55] Partly on the basis of the statements of "the interdimensional philosopher Seth," Grossinger constructed an inter- and transdimensional universe of which consciousness is the underlying constant.

Richard Grossinger is an example of a science-oriented mystic who developed transdimensional physics. While David Ash developed the physics of ascension, Richard Grossinger argued for an inter- and transdimensional physics, in which energies from higher dimensions transmit information to those people who are sensitive enough to download it. These people are called channels.

I will come back to many modern forms of channeling later. Some are reliable; others less so because these channels speak from personal convictions, which have to be released if the information is to be passed on cleanly. Pure information can therefore be passed on through pure channels—with an intention of unconditional love—for the greater whole. And that happens more and more these days. Why? I assume that the increase in vibration of the Earth has made the distance between the dimensions smaller. This makes information from those higher dimensions easier to download, provided it is received and passed on in purity. It is my understanding that we—humanity—are all on our way to becoming channels for higher dimensions. Eventually, we will go beyond those higher dimensions and the energies there, and thus, having become spiritually mature, we will only be in contact with our own higher self.

54. Grossinger, "An Unbottomable Void in Mind," 17.
55. Lorimer, "Review of *Bottoming Out the Universe*," 53–54.

The Older and Newer Ways of Channeling

Channeling has evolved greatly in the last hundred years. The techniques used in channeling are based on the shamanism of indigenous or natural peoples, ancient and contemporary.[56] This strong interest in shamanism and indigenous peoples is well known within the spiritual women's movement. The oldest technique rooted in shamanism is that of deep-trance mediumship. The method has been used by the oracle of Delphi and other major oracles and seers of the ancient world. In this case the medium is *not* aware of the information received. In the first two decades of the twentieth century virtually all channeling involved this type of deep trance and it was mainly aimed at the spirits of the dead.[57] Mrs. Blavatsky and her successors were strongly opposed to what they called "spiritism" in the late nineteenth and early twentieth centuries because they believed it would be dangerous and pointless.

However, there are also deep-trance mediums who attracted more highly evolved and ascended energies. An example is Jane Roberts, who from 1974 channeled the entity Seth while working in a deep trance. But deep-trance mediumship became increasingly rare throughout the twentieth century as other forms grew alongside it. This has to do with the increase in the vibrational frequency of the Earth, causing the veil between the dimensions to become thinner and thinner. To show the difference with modern forms, I would like to dwell first on this somewhat more old-fashioned form of channeling.

The following information was provided by Stuart Wilson, who went into regression and contacted the angel Alariel, the spokesman for a group of twelve angels who work together with the order of Melchizedek.[58] When performed by a trained medium, deep trance is an effective but artificial way to bypass the personality. A deep-trance

---

56. Greene, "The Celestial Ascent of the Soul," 5, 14; Wilson and Prentis, *Beyond Limitations*, 81, 87.
57. Wilson and Prentis, *Beyond Limitations*, 81.
58. Wilson and Prentis, *Beyond Limitations*, 4 with reference to *Power of the Magdalene*, 14 (about the first contact with Alariel), 196 (about the order of the Melchizedeks).

medium is kept in a state of consciousness that keeps out the direct influence of the personality. The personality is on hold. The consciousness of the medium is artificially narrowed and constrained to make this work possible. Changes take place in the body, making normal functioning from daytime consciousness impossible.

A deep-trance medium channels in a grounded manner, fully embodying the energy of the source of information. This gives the people who are present in the room the impression that the source is physically present. The transfer takes place slowly and consciously with the choice of words being dictated by the source. All of this takes place in a darkened room, often lit by a single candle. The channel sits with eyes closed during the séance.[59] In essence, this is an artificial process with several drawbacks. Special circumstances are required, including a quiet room with little light and a second person responsible for the safety of the medium. The medium should not be touched during the deep trance, nor should he or she be brought back from the deep trance too quickly. Unexpected physical contact can cause trauma and even long-term health problems.[60]

This method is now at the end of its development cycle, while the development of conscious channeling is only at the beginning.[61] Conscious channeling offers many benefits because the channel is continuously aware of what is happening and can immediately return to daytime consciousness to communicate with someone present in the room. It's a more natural, fluid, and easy process that doesn't require staying in a darkened room. In a conscious channeling, the master human soul or the angel or the star being is being channeled in a light way, which gives a fast and smooth transfer because a larger part of the consciousness of the channel cooperates. In addition, the source providing the information usually uses the channel's native language, although it remains difficult to express subtle concepts in modern languages. In addition, this resource must make use of the channel's vocabulary. If it

---

59. Wilson and Prentis, *Beyond Limitations*, 87.
60. Wilson and Prentis, *Beyond Limitations*, 84.
61. Wilson and Prentis, *Beyond Limitations*, 89.

is large, there are all kinds of options available to convey the ideas that want to be expressed. The source also uses the channel's mind structure: for example, if that person is a scientist, different concepts are available than if someone has qualities in another area.

The source thus has to deal with the following conditions: language, vocabulary, and the mind structure of the channel.[62] The channel uses their own words in modern channeling; the flow of ideas to be translated into words is faster, easier, and smoother. Usually the channel sits on a chair and keeps their eyes closed, but the room is not darkened.[63] To what extent the information the source wants to transfer is actually transferred depends on these circumstances, and it can never be a perfect transfer[64] because the source is not perfect either. The source is not omniscient, because omniscience does not exist on an individual level. Even the Elohim, the most highly individualized beings in the universe, do not claim to be omniscient. Here, too, there is free will and room for individual interpretation.

There are differences between deep-trance channeling and conscious channeling. In conscious channeling, the channel interprets the message and uses their own words to articulate the flow of information. He or she functions as a kind of interpreter/translator. In deep trance, the source has more control over the vocabulary, so some people prefer this method. But experience shows that channels or mediums who can completely put their personality aside are rare. And even a transmission of information in a deep trance is no guarantee for the complete reliability of the message. Often the medium itself is of good intention, but we do not know our own consciousness well enough to notice when the psyche moves in the ego direction and distorts the message. Yet the old technique has brought much to mankind; only now more emphasis is placed on conscious channeling.[65]

People will have to determine for themselves what is true for them

62. Wilson and Prentis, *Beyond Limitations*, 84–85.
63. Wilson and Prentis, *Beyond Limitations*, 88.
64. Wilson and Prentis, *Beyond Limitations*, 84–85.
65. Wilson and Prentis, *Beyond Limitations*, 87.

and what is not. Therefore the universe has built-in free will with the intent not to see humans as little children dependent on an omniscient source, but to bring them to spiritual maturity and adulthood.[66]

In conscious channeling, the channel simply switches to a different frequency of consciousness. First there is the flow of channeled information. Second there is the awareness of being connected to and united with the source, and third, the awareness as a spectator of coordinating the whole process and transferring it directly to the environment. A conscious channel can switch imperceptibly between these levels during a workshop and move fluidly within the energy of the source and then outward in their own personality. A therapist may use conscious channeling during a session with a client, and the client may not even notice it.[67]

As the frequency of the Earth increases, channeling becomes easier, provided the intent of the channel is completely pure and there is a firm will and conviction to serve the big picture. More people are showing up with an out-of-body experience, or with a near-death experience, or people who can see in subtle energy layers, or people who contact the deceased without danger and so on. All of this is because the Earth's frequency is increasing. This makes it easier for information to come through from higher dimensions. It is necessary, however, to develop your own discernment and not to take everything for granted.

The growth of human consciousness is reflected in the evolution of the channeling process. While it was a slow, laborious and uncomfortable process in the 1870s, the methods have become faster, lighter, more flexible and also more playful. That is how it will develop in the future. People can use it as a second opinion, a point of view in addition to their own perception. Also, there is the opportunity to expand awareness and explore new perspectives. It is a way to increase the flow of information between soul and personality, making the connection stronger. It can open new horizons and new ways of looking at the Earth. It can bring stronger inner guidance. But it should not be used as a substitute for

---

66. Wilson and Prentis, *Beyond Limitations*, 85.
67. Wilson and Prentis, *Beyond Limitations*, 88.

our own inner guidance. Your own inner guidance is paramount.[68]

Alariel explained that until 1930 hardly any angelic beings were channeled. But after 1930, angels became fashionable, resulting in a tidal wave of angel books and pictures. Since 1930, the angelic world has actively sought contact, so the channeling of angelic beings thereafter became an important element in the channeling of light beings.[69]

Alariel explained that instead of channeling master human souls, star beings, and angels, channeling in a subsequent stage of evolution means you can communicate with the totality of your being or achieve as much of that totality as you know. Practically speaking, this means channeling the energy and quality of your own soul, allowing your soul to speak through you and color your vision of life. When the soul illumines the personality, the ego confines itself to the function of the individual consciousness. It stops being master of the personality. Only then can you channel the highest aspect of your total being: your higher self.

Heightened Earth Frequencies

People have changed in these modern times as we collectively have higher vibrational levels. I also notice a process of change in myself in which my mind is more open to things that cannot be explained rationally. Personal changes are related to collective changes. The measurement of Earth's vibrational frequency has increased sharply since 2012 when, according to the Native American peoples such as the Maya and the Hopi, an epoch ended and a great new Platonic year of about twenty-six thousand years began. The Kemetic-Egyptian wisdom teacher Abd'El Hakim Awyan also spoke of an awakening, exactly like wisdom teachers from other indigenous peoples.[70] The Age of Aquarius announced itself.

Since then, the vibrational frequency of the Earth has increased by

---

68. Wilson and Prentis, *Beyond Limitations*, 90.
69. Wilson and Prentis, *Beyond Limitations*, 92.
70. Mehler, *The Land of Osiris*, 196; Mehler, *From Light into Darkness*, 17.

leaps and bounds, especially in 2015 and again in the spring of 2021 (see end of chapter 21 with reference to measurements of spiritual radiesthesia by the Dutch organization DJR-Advies). This means that humanity can easily connect with what the ancients called the Other World–the unseen, invisible dimensions that cannot be seen from a 3D perspective but can be felt and experienced. That makes ascending to higher dimensions easier. This also means that some channeled information can be included in a twenty-first century holistic new science in which there is room for subtle levels. I want to serve and develop that science from the perspective of spiritual women's studies, now that philosophical materialism is being buried.

A Short Introduction to the New Science

Professor Thomas S. Kuhn has been concerned with changes in science. The transition from a paradigm in crisis to a new one is not gradual. According to Kuhn, a paradigm shift takes place via a revolutionary breakthrough.[71] British author Anne Baring also speaks of a new paradigm and a scientific revolution. That revolution is currently underway with more and more scientists realizing that they are part of a living, conscious, and intelligent universe that is eternal.[72]

71. Thomas S. Kuhn, *The Structure of Scientific Revolutions*, Chicago: University of Chicago Press, 2012; Mehler, *The Land of Osiris*, xix, 31, 195.
72. Baring, *The Dream of the Cosmos*, 327–57; Betty Kovács, *Merchants of Light*, 34. The new science refreshes the relation between science and the shaman-mystic tradition.

2
WHAT THIS BOOK OFFERS

Four different sources are used in this book.

1. The biblical and early Christian canonical writings.
2. The apocryphal texts of the Old and the New Testaments. Many of the texts about the heavenly journey or ascension were banned, hidden, and labelled apocryphal, including the so-called Gnostic Christian texts. They contain in part the essence of the original Christianity.
3. The conventional historical writings from authors in classical antiquity.
4. Modern alternative sources, regression reports, and channeled information, including regression accounts from Dolores Cannon and the duo Stuart Wilson and Joanna Prentis, and the channeled information passed on by Anna, the mother of Mary and grandmother of Yeshua through Claire Heartsong.

It is clear from the book *The Essenes* by Stuart Wilson and Joanna Prentis that they value the information from the team who worked on the two books passed through Claire Heartsong titled *Anna, Grandmother of Jesus* and *Anna, The Voice of the Magdalenes*. Wilson and Prentis see both books as an important coproject.[1] Therefore, I

---

1. Wilson and Prentis, *The Magdalene Version*, 55.

decided to include information passed on through Heartsong among the alternative sources. Heartsong provides unique information about the childhood years of Yeshua and Miryam's family.

The big question is the common thread throughout this book. To what extent do the ancient sources confirm the alternative sources? Where do they differ, and where do the different sources support each other? To find out, I compare the information of ancient authors with the alternative sources. I apply the comparative method and ask: Do ancient authors' accounts match eyewitness accounts of people in regression? Do the latter add anything? Do we get a clearer, more complete and more human picture of that time? If so, then it is desirable not to exclude this material anymore.

There are several interludes within this book. In the interludes, the *historical* sources are separated from the *alternative* ones. The interludes are elaborations on themes that are discussed in the chapters. They are variations on a theme. If you want to follow the main line of the book, follow the chapters without reading the interludes, which provide theological and historical glimpses. They delve deeper into the backgrounds. To improve readability, the more detailed information has been placed in the separate interludes.

At the end of her *Gospel of the Beloved Companion*, Mary Magdalene shows the way to ascension. She is a wayshower. But what is ascension? Ascension has been understood as the path to the light, which is the process of making one's own energy system lighter. This has been experienced as a kind of "ascension" or "rising" into the light. The ancients spoke of a "heavenly journey." You made a heavenly ascent through the "spheres" or to use a modern word, "dimensions." You ascended to higher regions where you had a vision, then went back to Earth to explain the process. Compare the process of ascension—both past and present—to climbing a mountain in several stages. The higher you reach, the more ballast or luggage you shed. The higher you climb, the further the view and the deeper the insight. You emerge more and more in unconditional love and surrender to life.

Formerly the last stage of the way to the light was the final ascension or final absorption into the Kingdom of Heaven or the Kingdom of Light; this area was called the Fullness of Light or the Pleroma by Gnostic Christians. That final transition used to be possible only *after* death. The heavenly journey ended in a complete rapture into the light. Mary Magdalene described her own ascension process and her transition to the world of light. But this was not a one-way ascension. She did not die and return to Earth from this higher region; the visit was temporary. She had not fully ascended in her lifetime because she came back to her physical body.

In the twenty-first century, ascension is also about a rise to the light; you absorb more and more light in your own energy body and bring them into a higher vibration. This process is greatly facilitated today because the Earth is currently moving into a higher vibration. According to modern information, humanity can now make a collective ascension. The big difference with ascension in the past is that because of this much higher vibration of the Earth, you no longer have to die to fully ascend and merge into the fifth dimension. According to recent information, it is possible to convert the physical body into light already on Earth. The only condition is to feel unconditional love, trust and surrender to life, and open your heart. That is the core message that this book about ascension—following the example of Yeshua and Mary Magdalene—wants to convey.

In *The Gospel of the Beloved Companion*, written by Mary Magdalene (or Miryam as she calls herself), she ascends through seven branches and portals along the Tree of Life. It's like climbing a mountain in several stages. My book takes you through seven portals from individual ascension to collective ascension.

Portal One

In the first portal, five chapters deal with the immediate family of Yeshua and Mary Magdalene, an Essene family. They also detail the childhoods of the young Yeshua and Mary Magdalene before their public life began.

Portal Two

The second portal forms the heart of the book; based on twenty-first century information, insight is given into the collaboration between the adult Yeshua and Mary Magdalene during their public years. Chapter 9 discusses the role and unique contribution of the female disciples; they are arranged in six circles consisting of twelve women each. The first women's circle is presided over by Mary, the mother of Yeshua, with Mary Magdalene acting as a second leader. The biggest difference from the six men's circles is that the women formed a close team and, in contrast to the men, became completely committed to unconditional love and mutual trust. With this they supported the energy of love and light, which Yeshua and Mary Magdalene aimed to anchor together on Earth. This is necessary to turn the downward spiral of negative energy of revenge and violence on Earth into an upward spiral of love and forgiveness. Anchoring this positive energy on Earth is necessary because otherwise this high-vibrating energy will not be able to reach Earth, which at that time vibrated at a very low frequency.

Portal Three

The chapters in the third portal take a closer look at *The Gospel of the Beloved Companion* written by Mary Magdalene herself. Later she took it with her to the south of France, where it was handed down within her Hebrew-French descendants until it came into the possession of the Cathars in the twelfth century. After an introductory chapter about this gospel, the apotheosis follows. Mary Magdalene, who calls herself Miryam in her own text (hence I also speak of the Gospel of Miryam), describes her own ascension process along the Tree of Life.

When Miryam passes through the first four levels or portals, she traverses the third and fourth dimension and reaches the fifth dimension. Miryam surprisingly finds out that here there is no duality anymore. She has transcended it, just as she has transcended the master of the world, who tries with all his might to stop her at the transition from the fourth to the fifth plane. Her transit is successful. From the eighth level she reaches the Fullness of Light or the Kingdom of Heaven. Here a lady, shining with brilliant light, awaits her with open arms.

Surprisingly God the Mother, Spirit, is present in Miryam's ascension experience. Her other name is Wisdom. Wisdom was loved and venerated in the First Temple tradition but had to leave when patriarchal mono-Yahwism and monotheism that became popular after the destruction of the First Temple in 586 BCE, definitively won the battle between the old Hebrew religion and what was called "the Reformation" of it. From the sixth century BCE onward Wisdom had to go into hiding. For a while she became a Lost Lady, but she succeeded to live on in groups who left Jerusalem and fled abroad, and the Essenes were one of these groups. In Miryam's gnostic text the Lost Lady returns. It is the Mother (Wisdom and her ancient Wisdom tradition) that invokes and stimulates the individual and collective ascension process. With the Mother, we temporarily lost the knowledge of ascension. Miryam brings her back to activate this knowledge in our modern consciousness. She shows the way.

In Miryam's *Gospel of the Beloved Companion*, Miryam is appointed as leader and successor by Yeshua at the Last Supper. However, her leadership is not accepted by a certain group of patriarchally oriented male disciples led by Peter, though her leadership is accepted by those male disciples who have an Essene background, as well as the female disciples. After the abolition of the Essene order, these merged into the groups of the progressive gnostics. These liberal gnostic groups have no problem with female leadership. Mary Magdalene develops into the leader of the Gnostic Christian mystery school, in which the process of ascension is central.

Portal Four

In this portal, the book turns to Yeshua. Did he survive the crucifixion and travel to Central Asia? Historical sources are introduced that shed an unexpected light on this issue.

Portal Five

Here I discuss a key figure in the crucifixion and resurrection: Joseph of Arimathea, who invented smoke screens around the disappearance of Yeshua—as well as Mary Anna and Mary Magdalene—after the resurrection. This mysterious key figure deserves more attention in the twenty-first century historiography of early Christianity. We'll explore his role in these important events as well as the economic and societal factors that allowed him to provide safe transport for himself and his family when he came under suspicion by the Pharisees.

Portal Six

The book follows the footsteps of Mary Magdalene in the south of France and England. Are there historical sources that show she had been here? Are they reliable? Can we trace her trail from antiquity to the Middle Ages? The answer is: Yes. I found seven links to show how Miryam's legacy was passed on by her Hebrew-French descendants to the medieval French church and the French Cathars. I also discuss a text smuggled out of the besieged Cathar stronghold of Montségur Castle in March 1244—could it be the ancient gnostic text of *The Gospel of the Beloved Companion*? Could Miryam's Gospel be the link between the actions of Mary Magdalene in the south of France in ancient times and the Cathars in the Middle Ages?

Portal Seven

The last chapters focus on the phenomenon of ascension in the twentieth and twenty-first centuries. The concept of ascension is lifted out of the religious-mystical context into which it has been placed for so long. In the twentieth and twenty-first centuries, quantum physics started to take off, which makes it possible to see ascension as a natural phenomenon and to provide a scientific explanation for it. This development synchronizes with the revolution in consciousness that has been taking place on Earth in various stages since 1875. Key moments are 2012 and 2021. The increasing vibrations of the Earth and her inhabitants with opened hearts seem to be leading to a collective ascension.

Summary

At the end of the search, we are able to provide answers to the key questions. Are the alternative sources supported by historical sources? Do they broaden our field of vision? Can we supplement traditional history with the contributions of Mary, the mother of Jesus, Mary Magdalene, his wife, and the many female disciples? Will this create a more balanced picture, which better suits our modern times?

PORTAL ONE

THE ESSENE FAMILY OF MIRYAM AND YESHUA

Figure 3.0. Leonardo da Vinci, *The Virgin and Child with Saint Anne*, c. 1501–1519. Louvre, Paris.

THE HISTORICAL AND ALTERNATIVE SOURCES

The Gospel of the Beloved Companion by Miryam the Migdalah did not give extensive information about the family of Yeshua and Miryam and their possible Essene background. Miryam restricted herself in her Gospel to a description of the highlights of the public years. During this time, as the companion and beloved of Jesus, she acted as his coworker. Portal One is about gaining insight into the family of Miryam and Yeshua, with the key question being: Are they an Essene family?

Little is known about Miryam and Yeshua's family in conventional historical sources. The scarce sources that are available provide both similarities and different information here and there. It's difficult to get any further because too many puzzle pieces are missing. Jehanne de Quillan, who translated *The Gospel of the Beloved Companion* from the ancient Greek into French and English, made a few remarks in her book about the fact that recent research showed the role of the Essenes was a supporting and connecting organization. But these were no more than allusions. The following question is now on the table: What information do traditional and alternative sources provide about the possible Essene background of the families from which Miryam and Yeshua descended?

Each of the following chapters investigates the canonical, apocryphal, and historical sources as well as alternative sources (regression reports and channeled material) using the interdisciplinary and comparative method.

3

ANNA, MOTHER OF MARY

Anna is the mother of Mary and grandmother of Jesus. I examine Anna from different angles, including: biblical and apocryphal traditions, the pre-Christian mother goddess Ana in the Celtic tradition, the Christian Anna, and Anna according to alternative sources.

The name Anna did not appear in the canonical biblical sources. We came across the name Hannah, a widow in the temple who recognized Jesus as the Messiah.[1] However, the name Anna was mentioned in an apocryphal source.

Anna appeared in one of the oldest extra-biblical texts to the New Testament: *The Birth of Mary*, later called *Infancy Gospel of James*, also known as the *Protevangelium of James*. It is a text that is becoming extremely popular within Western and Eastern Christianity. The story goes as follows: Anna was barren, yet she conceived of the Lord and not of her then-absent husband Joachim. She had a daughter Mary who also became pregnant from the Lord without the intervention of her husband Joseph. After the birth of Jesus, midwife Salome experimentally established that Mary was still a virgin. There are still many beautiful details to report about this extra-biblical text, but I will limit myself here to the information about Anna. It is known that in the Eastern and Western churches from the fourth century and especially from the fifteenth century that people were increasingly convinced that Mary was born "immaculate" of Anna and that she therefore gave birth to an

---

1. Lk. 2:36–38.

immaculate Jesus. In later church dogmatics, "immaculate" meant not to be burdened with original sin.

Anna's becoming pregnant without any physical intervention by a man remarkably ties in with the information from alternative sources about light conception, both in the regression reports and in Claire Heartsong's channeled book about Anna, which states that after Anna had been single for a long period in her life, she married. Then the conception of Mary was the result of a light conception without physical intervention by Anna's husband Joachim. Another parallel is that in *The Birth of Mary*, Anna promised to dedicate her child to the temple at the age of three. In the channeled message, Anna and Joachim handed their three-year-old daughter Mary over, but not to the temple in Jerusalem, but the temple on Mount Carmel. People of the mystery school there received her and took care of her further education.

The Celtic Mother Goddess Ana

The name of the great goddess of the Celts is Ana, otherwise known as Anu or Dana. She was the ancient Earth Mother of European peoples from prehistoric times before the Celts. The name Ana means "Mother." She is a manifestation of the Primeval Mother or Great Mother. She shows herself in three aspects: the white of early spring, the red of full summer, and the black of autumn and winter; from the black the new white light is born. When the Celts arrived, she took a male partner, Lug, also known as Dagda or Bran. Ana remained popular in Celtic Ireland, England, and Brittany. She was especially honored in her grandmother aspect, her black aspect.[2] She is the goddess of the cave and the waters. She is the goddess of death who regenerates life, who carries ancestral souls into newborns. She brings nature and man back to life in eternal cycles and spirals.[3]

---

2. Marija Gimbutas, *The Civilization of the Goddess*, San Francisco: HarperCollins, 1991, 305; van der Meer, *The Black Madonna from Primal to Final Times*, 184n14.
3. Marija Gimbutas, *The Living Goddesses*, ed. Miriam Robbins Dexter, Berkeley: University of California Press, 2001, 185; Begg, *The Cult of the Black Virgin*, 85; van der Meer, *The Black Madonna from Primal to Final Times*, 184n15.

From the Mother Goddess Ana to the Christian Saint Anna, Grandmother of Jesus

In the French coastal region of Brittany, people love the mother goddess Ana.[4] There the roots of the Christian Saint Anna arguably go back to the worship of the Celtic mother goddess Ana; elsewhere that process is more difficult to follow. In Brittany and in surrounding areas, such as in the Netherlands and Germany, a great veneration for Saint Anna flourished at the end of the Middle Ages.[5] The goddess Ana transformed in Christianity into Saint Anna, the mother of Mary and the grandmother of Jesus.

Was Anna a Celtic Princess from Britannia? The Historical Sources

In *The Black Madonna from Primal to Final Times*, I established that Anna was quite popular in Brittany in France without being able to give a specific reason for this.[6] Later I read that according to an old tradition Anna may have been a Celtic princess.[7] I took that for granted in 2018, but actually, I couldn't believe that the copper-colored hair and blue eyes that Jesus has in some apocryphal texts would come from a Celtic grandmother. Then, I came into contact with the alternative sources. Anna of Carmel came through Claire Heartsong's transmitted book that she had been to Britannia (Britain) several times in her long life and was adopted there as the daughter of a royal family. Anna traveled, both by sea and by land route, through present-day France and the ports in Brittany, to Avalon in present-day England. Researching the historical sources for this book, Anna's Celtic origin again popped up. Several British authors have

---

4. Jean Markale, *Cathedral of the Black Madonna: The Druids and the Mysteries of Chartres*, Rochester, VT: Inner Traditions, 2004, 151; van der Meer, *The Black Madonna from Primal to Final Times*, 184n16.

5. van der Meer, *The Black Madonna from Primal to Final Times*, 185n17.

6. van der Meer, *The Black Madonna from Primal to Final Times*, 184–85.

7. van Dijk, *Maria Magdalena, de Lady van Glastonbury en Iona*, 107.

pointed out that there has been a connection between the House of David and the Royal Houses of Britain.[8] Barry Dunford pointed to an ancient tradition that Anna, Mary's mother, was originally from the Cornish royal house.[9] Against this background, the many legends in Brittany and Britain about Anna and the love of the people for her become more understandable (chapters 14 and 16).

The Regression Reports

Joanna Prentis and Stuart Wilson were acquainted with the traditional sources that identified Anna as a Celtic princess. They relied among others on Lionel Smithett Lewis.[10] He stated that according to a Breton tradition from France, Anna was from Cornwall and of royal blood. Wilson and Prentis decided to get back to Daniel Benezra. In a regression session, Prentis asked: "What do you know about Mary's mother and your friend Joseph of Arimathea?"

Daniel replied, "She was a very special person, a princess from a Celtic family in England. She was a very wise and gracious person, but also someone of great authority. We recognized her as one of great powers, a high initiate. The name of Joseph's mother (here he means Joseph of Arimathea) was Anna, but in the family, this was pronounced Ayna."[11]

Daniel also said that Anna went back to England after Mary's marriage to Joseph and died there. He said, "I always knew she was an advanced being, and I was not surprised when I asked about her in the Interlife. I was told that she had made her ascension towards the end of her life."[12]

That there is a possible bond between Anna and Britannia will come back in the following channeled messages about Anna, but in a different way than Daniel suggested.

---

8. E. Raymond Capt, George F. Jowett, Robert Mock, and Isabel Hill Elder point to this connection; see van Dijk, *Maria Magdalena, de lady van Glastonbury and Iona*, 106.
9. Dunford, *Vision of Albion*, 94.
10. Lewis, *St. Joseph of Arimathea at Glastonbury*, 63; Wilson and Prentis, *The Essenes*, 229.
11. Wilson and Prentis, *The Essenes*, 230.
12. Wilson and Prentis, *The Essenes*, 232.

Anna of Carmel in the Channeled Sources

In the messages channeled by Claire Heartsong, Anna introduced herself as Anna of Carmel.[13] The Carmel Mountains are a mountain range in northwestern Israel, right on the coast (fig. 5.1, map). Mount Carmel is the northernmost peak of that mountain range and has a height of approximately five hundred meters. Elijah's stone is located there; this is what Anna called the altar that the prophet Elijah built there shortly before he made his heavenly ascent in the chariot of the sun.[14] According to Anna's information, Mount Carmel had a thriving Essene community of women, men, and children with a school and library. This agrees with certain modern scientific insights[15] and also with the regression reports, which list Mount Carmel as one of three northern Essene settlements in Palestine.[16]

In the foreword to Claire Heartsong's first book it is written that between 1930 and 1940 the American clairvoyant Edgar Cayce, during a self-induced trance state in which he gave 14,000 readings, gave "the first fascinating description" of the Essene community on Mount Carmel.[17] Cayce was convinced of a connection between Jesus's family and friends and the Essenes. He saw the Essenes as successors to the teachings of Melchizedek, as expounded and proclaimed by the prophets Elijah, Elisha, and Samuel. Cayce made a connection between the ancient wisdom teachings and the school of prophets on Mount Carmel from which the Essene community later emerged.[18] The view that the Essenes were the successors of the Melchizedek teachings and the heirs of a school of prophets around Isaiah in Judea is supported by historical evidence from recent scientific research.

---

13. Heartsong, *Anna, Grandmother of Jesus*, 122.
14. Heartsong, *Anna, Grandmother of Jesus*, 110 (about Elijah); 41 (this is the place where the ascension of Hismariam takes place); 95 (on the top of the mountain Anne receives a vision).
15. Goranson, "On the Hypothesis that Essenes Lived on Mt Karmel," 563–567.
16. Wilson and Prentis, *The Essenes*, 39, 160.
17. Heartsong, *Anna, Grandmother of Jesus*, xvi.
18. Heartsong, *Anna, Grandmother of Jesus*, xvi, xvii, xix.

Cayce described Anna as a visionary and prophetess who was held in high esteem by the Essenes. She initiated an innumerable number of people, including twelve girls whose purity was deemed high enough for them to be candidates for an important event to come, such as bringing the Messiah or the Anointed One of the Light into the world through a light conception—much like Anna giving birth to Mary through light conception and Mary doing the same with Yeshua. Mary Magdalene was also conceived in this way, as well as many others from Anna's extended family. Anna was the ancestor of both families: the family of Miryam of Bethany and the family of Yeshua.[19]

Anna stated that she lived to be about six hundred years old. Before reading this, I had dived into books by Romanian author Radu Cinamar, in which he explained esoteric knowledge in a rational way.[20] He also talked about the possibility that more advanced people can extend their lives through inner technology. He described modern people who were able to stay young for many years, which means they do not have to be born again from the mother's womb every time. Sometimes they take certain highly secret substances to prolong their lives; sometimes they achieved a long life by withdrawing for extended periods into complete meditation where they rest. You can now understand where all those science fiction films come from. It is another reality that is presented as imaginary or impossible, but it is indeed possible. Anna reported that she sometimes rested while bandaged for months and even years in caves maintained by the Essene brotherhood and sisterhood near Mount Carmel and Hermon and in a certain cave near Qumran.[21]

Radu Cinamar also wrote about groups of human beings who lived in the inner earth at different energy levels.[22] Anna and a certain group of Essene brothers and sisters had done this for many years in Egypt. I

19. Heartsong, Anna, Grandmother of Jesus, xvii. Anne is the holy woman who is the pillar and source of inspiration behind the mission of the Essenes.
20. Cinamar, Transylvanian Moonrise, 31–81 discusses the mindblowing story of Elinor.
21. Heartsong, Anna, Grandmother of Jesus, 47.
22. Cinamar, Inside the Earth: The Second Tunnel. Chapter 3 discusses the city of Tomassis. Chapter 4 informs about the crystal city of Apellos in the inner earth.

was lucky enough to have learned about all of this before I became serious about Anna's story as channeled by Claire.[23]

Anna's early years shed light to the story. She was born as Hanna in 612 BCE in Etam, a village south of Bethlehem in Judea. She died in England in 62 CE at the age of about six hundred years old.[24] In 597 BCE, Anna asked the Councils of Light if she may return to Earth. It was shortly before the time when the Babylonian conquerors swept through the village of Etam near Bethlehem and occupied Jerusalem a second time and before Hanna's wedding with Tomas, and she was pregnant by him. Tomas was captured by the conquerors and taken away as an exile, as were Hanna's parents. They left behind Hanna, descendant of the house of David, who was forced to watch as her two brothers were killed before her eyes, and she was also raped by the group. She had a near-death experience, and Anna's soul was allowed to descend into Hanna's body and connect with Hanna's soul.[25] So, when Anna's soul descended into Hanna, she was a "walk-in." With the help of Naomi, Anna (as Hanna) gave birth to the daughter Hanna was pregnant with, Aurianna, on May 23, 596 BCE. For thirteen years, Anna and her daughter worked as midwives around Bethlehem. Later, Anna was taken care of in Jerusalem in the house of Hanna Elizabeth and Johannes, both Essenes. Johannes, who was a writer, became Anna's tutor.

Later, Aurianna married an Essene from Carmel. Anna felt the inner call to enter the Carmel mystery school, "one of the oldest mystery schools that survived the rise of many civilizations."[26] Aurianna had children and a granddaughter called Hismariam. Hismariam made her ascension around 150 BCE on Mount Carmel and later incarnated as Mary Anna, the mother of Yeshua.

When Anna was 39 years old, she went overland to Egypt, where

23. Heartsong, *Anna, Grandmother of Jesus*, 18, 37, 47.

24. Heartsong, *Anna, Grandmother of Jesus*, 284, Chronology Chart, 288. Heartsong and Clemett, *Anna, The Voice of the Magdalenes*, 374: the year 82 is given as Anna's year of death and burial in the Tor; 379: Anna's life is expanded from 634 BCE to CE 82.

25. Heartsong, *Anna, Grandmother of Jesus*, 9–11.

26. Heartsong, *Anna, Grandmother of Jesus*, 16.

she stayed for 303 years. She first stayed in the ruins of the temple complex of On of Heliopolis where many Hebrews lived in small houses around the temple. This is confirmed by historical evidence.[27] She lived in the underground city of Tat, whose subterranean passages stretched from the Great Pyramid to the harbor "which you call Alexandria."[28] From there they had access to a network in the inner Earth; some people speak of Agartha. Anna was employed copying scrolls, among other things. Forty years after arriving in Egypt, she became an initiated priestess of Isis, Hathor, and Sekhmet. In addition to her work as a writer and copyist, she knew how to bilocate—to be in two places at once and teleport or move herself from one place to another in the blink of an eye. She knew the Isis mystery of resurrection in which the cells are rejuvenated into a form of eternal youth. Anna was now known as the high priestess of the Great Mother.

After this, Anna prepared for a future departure to Palestine. She moved to the port city of Alexandria for her last thirty years in Egypt. Again, she went underground with other Essenes for the first nine years. They had brought numerous copies from On that needed to be copied. As writers, they had access to the above-ground library of Alexandria. She regularly visited the temple of Isis above ground and was admitted there as a priestess. During her last hundred years in Egypt Anna regularly traveled to England, Greece, southern Gaul, and other Mediterranean countries. She felt especially at home in England, where she was adopted as one of the daughters of a Celtic leader and received a small tattoo by which she was recognized many years later. She foresaw that she would consciously "pass through the veil" and pass over in this land. On her return to Egypt she found that many Hebrew people now lived there, which caused tensions with the Greek inhabitants of Egypt. Anna left Egypt in 207 BCE, with her great-granddaughter Hismariam, who had traveled with her from Alexandria.

Anna and her Essene group arrived in the port city known as Akko,

27. Josephus, *Antiquities*, 13.3.1; Barker, *The Mother of the Lord*, 12n24.
28. Heartsong, *Anna, Grandmother of Jesus*, 20.

north of Mount Carmel. Anna said that for more than two centuries (207 BCE to 28 CE), she spread the wisdom teachings, initiated people, collected and cultivated medicinal herbs, and further developed the community. As a high priestess in the Essene order, she visited the villages and communities in the surrounding area together with others in her physical body. She also used bilocation and teleportation to communicate with people who were further away and regularly retreated to caves during these excursions to maintain her physical body. There are graves that are filled with a womb-like fluid, and in her own words, in this way she avoided having to incarnate again as a baby or descend again as a "walk-in" into someone's body.

Although she was single and lived celibate for a long time, Anna felt the need to get married. In December 58 BCE she married the Essene Matthias. A light conception followed, and Anna became pregnant with her eldest child and first son, Joseph, later Joseph of Arimathea. He was born "according to your Gregorian calendar" in 57 BCE. The birth of a daughter, Martha, followed a year and a half later in 55 BCE. She was steadfast, decisive, wise, and had a knack for managing details. Joseph had a daughter in a second marriage via a light birth: Mary of Bethany later called Mary the Migdalah or Mary Magdalene, which also makes Anna the grandmother of Mary Magdalene. Anna's husband Matthias preferred celibacy and moved to the community in Qumran. The marriage was dissolved in time. Matthias was wounded and died after the Roman legions burned down Qumran at King Herod's instigation. Anna was left with two small children and felt a strong urge to marry again, and she had a vision in which she gave birth to another twelve children.

In the late summer of the year 52 BCE, she was tending to the plants in the garden at daybreak when her inner ear heard a kind of sound similar to that of a bagpipe. She went out and climbed the mountain, and toward the evening she saw and heard how, somewhere down below, a minstrel leaned against a cedar while playing the bagpipes and the lyre. In Joachim she recognized her soul mate. Three months after the betrothal, he felt the call to go to India in the early spring, only to return after two years in the late autumn of 49 BCE. A month later, they were married.

In a period of sixteen years, eleven children were born: Ruth; twins Isaac and Andreas; twins Mariamne and Jacob; Josephus; twins Nathan and Lucas; and after two miscarriages, Rebekah, Ezekiel, and Noah were born successively.[29] The archangel Gabriel announced a twelfth birth. He also told her that in this conception, Joachim's physical seed will not reach Anna's womb. However, on an ethereal level, his seed and essence will mix with her essence. Although Anna prepared for the foreseen conception, nothing happened for many years. Anna explained that this time was needed to be able to reach and hold the intense light frequency necessary for Mary Anna's conception; otherwise, it would have resulted in a miscarriage.[30]

Her eldest son Joseph of Arimathea invited Anna and her husband Joachim for an upcoming trip to England. Joseph took his first wife, Eunice Salome with him; she was in poor health after the birth of two daughters. Their two daughters stayed at home. They set course for Alexandria where they stayed for two months, after which with a convoy of three cargo ships they went to Marseille, where new cargo was loaded. From there they went to the Languedoc to stay with an Essene community at the foot of the Pyrenees. Anna's sons Jacob and Isaac, and Isaac's wife Tabitha stayed behind here.[31] They continued through the Strait of Gibraltar across the Atlantic Ocean to Brittany, to the island of Mont Saint-Michel. After this, they crossed the Channel and entered the west coast of Cornwall, just before great storm winds arose and rain poured down.

Via the Brue River they reached the island of Avalon or Apple Island, also called *Ynys Witrin* or mystic island (see chapter 13). It is now known as Glastonbury or the Isle of Glass, Anna said. Everything has remained exactly as Anna found it around 300 BCE. They visited

29. Heartsong, *Anna, Grandmother of Jesus*, 77.
30. Heartsong, *Anna, Grandmother of Jesus*, 79.
31. Heartsong, *Anna, Grandmother of Jesus*, 82.

an island off the coast of present-day Wales, also called Avalon or Mona by the ancients and known as the Isle of Anglesey. All kinds of excursions were made, including to Stonehenge and Avebury.[32]

Anna was recognized by the Celtic tribe who adopted her long ago as the one who had come from Egypt to England. They recognized her by her small tattoo in the shape of a trident near the hairline on her forehead. She was recognized as the Druidic high priestess who carried the traditional knowledge and energies of the Great Mother. She again wrapped herself in the white robes of the Druid order.[33] During a visit to the stone circles on Mona or the present-day Isle of Anglesey, where the Druids have an important shrine, the Archangel Gabriel visited Anna again. The incarnation process of the expected one was announced, but the actual conception would follow later.

Anna and Joachim made their way to Cornwall, where numerous additional freighters were being prepared for the return voyage to Palestine. Three of her children were left behind in southern England: Andreas, age 26; Josephus, age 21; and Noah, age 12. They stayed behind on the islands of Glastonbury and Mona in order to receive their initiations with the druids. Anna and Joachim traveled back with Joseph of Arimathea and his wife Eunice, but first they visited the Languedoc again to stay with relatives there for a month, because Isaac and Tabitha had settled at Mount Bugarach. Next, they traveled back across the Mediterranean Sea to Ephesus on the west coast of present-day Turkey. The situation in Palestine was troubled because of King Herod's harsh edicts, his immoral behavior, and his insensitivity to the needs of the people. Carmel and Qumran were no longer the safe havens they once were. It was decided to evacuate part of the inhabitants of Carmel and to dismantle part of the library. This time it was not transferred to Qumran but to Ephesus. There, the scrolls were scattered in inaccessible caves in isolated places along the coast. Anna decided to stay in Ephesus to await the birth of her twelfth child and to help dis-

---

32. Heartsong, *Anna, Grandmother of Jesus*, 83.
33. Heartsong, *Anna, Grandmother of Jesus*, 84.

tribute the scrolls with texts that she herself wrote about three hundred years previously. Some of these communities became the basis for later Christian monasteries.[34]

On a late December night in the year 21 BCE, one month after their arrival in Ephesus, the archangel Gabriel visited again. Anna and Joachim both ascended high and met the one who would be called Mary Anna in a coming incarnation. Anna knew her as the incarnation of Hismariam, one of the descendants of Hanna's/Anna's daughter, with whom Anna felt quite close during her long life.[35]

Anna and Joachim felt Mary Anna's presence in every cell of their bodies. Joachim also perceived her light and was completely absorbed in it. Anna explained to Heartsong that Mary Anna was conceived differently from the other eleven children. Anna thus shared that this light conception took place without fertilization on a physical, but one on an etheric level, with Mary Anna descending into her womb to fulfill the promise of a virgin conception and birth.[36]

After a blessed pregnancy without illness or suffering, a painless delivery followed.[37] It was known to all in Anna's circle that this little baby girl, Mary Anna, would give birth to the Anointed One, Yeshua, whose coming was foretold in the prophecies. Thus, the high priestess Anna not only became the mother of Mary Anna, but in time also the grandmother of Yeshua. Anna's long life was all about preparing for this role.[38]

Later on Mount Carmel, Anna began to prepare her grandson Yeshua for the great initiation of the crucifixion and resurrection. She took him to caves and initiated him. She had a clear task in his education.[39] When he was between twenty-nine and thirty years old,

---

34. Heartsong, *Anna, Grandmother of Jesus*, 85–86.

35. Heartsong, *Anna, Grandmother of Jesus*, 39 (She was previously incarnated as Tiye, the daughter of Joseph the Israelite, who was sold to Egypt. Tiye becomes the wife of Pharaoh Amenhotep III and the mother of Amenhotep IV aka Pharaoh Akhenaten), 86.

36. Heartsong, *Anna, Grandmother of Jesus*, 87.

37. Heartsong, *Anna, Grandmother of Jesus*, 87–88.

38. Heartsong, *Anna, Grandmother of Jesus*, 87.

39. Heartsong, *Anna, Grandmother of Jesus*, 145.

she undertook journeys with Yeshua.[40] And after Yeshua's crucifixion, which took place in Anna's chronology in 30 CE, Anna and her family fled Palestine in 32 CE and moved to the south of France.[41]

During her stay in the south of France, Anna also visited England again; she wanted to die there and be buried on the Tor[42] (see chapter 16 and the photos of the Tor in fig. 16.2a–b). She liked the green and sloping land, and she was known as the adopted daughter of a Celtic tribe, but also she had family there who took her in—the three children she'd left behind around 22 BCE when she and Joachim left for Palestine.[43]

For now, the conclusion must be that there is historical evidence showing how there was an Essene community in Heliopolis around the ancient temple of On and on Mount Carmel, as well as some legends that imply a link between Anna and England.

Together with all the children, grandchildren, and great-grandchildren of grandmother Anna, a close-knit extended family had emerged. This family formed the support network for the mission of Yeshua and Mary Magdalene, who were both grandchildren of Anna. In that large family, which spread across the world after the crucifixion, they all played important roles. But there was one who played a key role in supporting Anna, Mary Anna, Yeshua, and Mary Magdalene: Joseph of Arimathea. I invite you to follow the trail of Anna's first child and eldest son, Joseph, with me—the trail that was prepared by Anna of Carmel.

---

40. Heartsong, *Anna, Grandmother of Jesus*, 288.

41. Heartsong, *Anna, The Voice of the Magdalenes*, 368. In the spring of the year 32 Joseph took his mother Anna and their family to Alexandria, and from there they traveled to Saintes-Maries-de-le-Mer, arriving in the middle of summer. By the beginning of October they arrived at Mount Bugarach and settled there. In the late autumn of 38 CE, Anna and part of her family went with Joseph to Avalon in England.

42. Heartsong, *Anna, The Voice of the Magdalenes*, 374.

43. Heartsong, *Anna, Grandmother of Jesus*, 85.

4

JOSEPH OF ARIMATHEA

Very little is mentioned about Joseph of Arimathea in the Bible. There is only brief information about Joseph given by the evangelists at the end of their accounts, not earlier.[1] If you put all these separate fragments together, the following picture emerges.

Joseph was a rich man. He was a man of distinction and a member of the Sanhedrin, the highest Jewish legislative and judicial Jewish court. He was secretly a disciple of Jesus; he established contact with Pilate and cooperated with Nicodemus to bury Jesus's body in the tomb. But was Joseph related to Yeshua or not? It is noted that in the Jewish tradition, only close relatives are involved in a funeral. Western tradition is silent about any family relationship between Joseph and Jesus. In the Eastern tradition, on the other hand, Joseph was indeed a relative of Jesus, but the Bible says nothing about that.

Joseph appears more often in the apocryphal texts than in the biblical sources, as summarized below.

- In *The Gospel of the Beloved Companion* written by Mary Magdalene or Miryam herself, she reported—as the beloved disciple, alias John, does in his gospel—that Joseph of Arimathea, "a disciple of Jesus, but secretly for fear of the Pharisees," went to Pilate and obtained permission to take the body. Joseph

---

1. Mt. 27:57–61; Mk. 15:43; Jn. 19:38 and Jn. 19:39–42.

cooperated with Nicodemus, who took many herbs with him.[2] Subsequently, there is a difference with the Gospel of John. Not Joseph and Nicodemus, but Miryam the Migdalah and Miryam the mother of Yeshua took care of his damaged body. They bound it with the herbs in linen cloths, "as the burial rite demands."[3] Within the Jewish tradition, the immediate female relatives cared for the body of the deceased. Miryam's version can therefore be regarded as the most reliable. In Miryam's text, the women are fully involved and have not yet been written out, as is the case in the canonical texts. When Miryam discovered that the tomb was empty on the Sunday morning and met Yeshua at the tomb, she returned to her home in Bethany to tell the good news. Among the small group of relatives and students—most of the students had fled—is Joseph of Arimathea.[4]

- The Gospel of Peter was one of the oldest apocryphal gospels. Here Joseph was mentioned in relation to Pilate.[5]
- The Gospel of Nicodemus gave more detail about Joseph, "a member of the High Council, from the village of Arimathea, who himself also expected the kingdom of God."[6] After he went to Pilate and received the body of Jesus, the Jews were enraged and seized Joseph, locked him up, sealed the door, and set a guard before it. When they went to pick up Joseph after the Sabbath, the door was still sealed but Joseph had gone—he'd escaped. It is clear from this text that Joseph openly incurred the enmity of the Pharisees and he was imprisoned. This information partly agrees with the regression reports, which state the Pharisees were furious about the role that Joseph played in removing Jesus from the cross. They now understood that he, too, was an Essene, which put him

---

2. *GBC* [40:1], 74; *MMU*, 455.
3. *GBC* [40:2], 74; *MMU*, 456.
4. *GBC* [40:7], 75; *MMU*, 465.
5. The Gospel of Peter, 1.3–5a. See Klijn, *Apocriefen van het Nieuwe Testament* I, 32.
6. The Gospel of Nicodemus 12–14. See Klijn, *Apocriefen van het Nieuwe Testament* I, 57–90, 70–73.

into immediate danger. More information is provided about this subject in chapters 12 and 13.

The Jewish Sources

In the Jewish sources and according to the Talmud, Joseph of Arimathea was a younger brother of Joachim (who was the father of Yeshua's mother, Mary).

- *Joseph of Arimathea, an uncle of Mary.* Joseph was therefore Mary's uncle and Jesus's great-uncle. After the death of Mary's husband, also called Joseph, Joseph of Arimathea took over the care of the family. This family relationship made him the legal father of Jesus after Joseph's death.[7] This family relationship also implied that he was descended from the house of David. The Eastern Orthodox tradition also held that Joseph of Arimathea is the great-uncle of Jesus.
- *Joseph of Arimathea, a much older brother of Mary.* Other people, who were also convinced of a family relationship, believed that he was a much older brother of Mary and therefore an uncle of Jesus. This last vision can be found in Daniel Benezra's vision in the regression reports and in the channeled messages that Anna passed on through Claire Heartsong. Stuart Wilson, in a commentary, believed that Joseph was Mary's older brother. Claire Heartsong also saw it that way and mentioned an age difference of thirty-seven years.[8] After the crucifixion Joseph went to England where he was active. (See the English Interlude and the Celtic Interlude.)

---

7. Jowett, *The Drama of the Lost Disciples*, 18; Prophet, *Mary Magdalene and the Divine Feminine*, 29, 251.

8. Wilson and Prentis, *The Essenes*, 212; Heartsong, *Anna, the Voice of the Magdalenes*, appendix C, 379; Cannon, *Jesus and the Essenes*, 221. Suddi sees it differently; he mentions Joseph as a cousin of Jesus's mother; he also states that Joseph undertakes travels with Jesus and that sometimes they are accompanied by Jesus's mother.

In any case, Joseph must have been one of the richest and most influential men of his time: you could call him the Onassis of antiquity. The great wealth had to do with his dominant position in the tin trade. Tin was the major component of bronze, and bronze was vital to the military apparatus of the Romans (see a Tin Interlude and a Second Tin Interlude).

The Channeled Sources

In the channeled sources and especially in the family tree provided by Claire Heartsong, Joseph of Arimathea (57 BCE–62 CE) was the eldest son of Anna from her first marriage to Matthias. According to the channeled information, Joseph was therefore a half-brother of his sister Mary Anna (20 BCE–66 CE), who, as the youngest of twelve children born to Anna during her second marriage, was thirty-seven years his junior. So, Joseph was the head of the family because as the oldest brother, he took care of his much younger sister and her children after the death of Mary Anna's husband, Joseph (37 BCE–20 CE).

In Claire Heartsong's first book she channeled Anna, who described Joseph's childhood. Anna announced that a gifted son was born from her first marriage to Matthias and within the Essene community on Mount Carmel. This eldest son was predicted to play a key role in the lives of Anna's youngest daughter, Mary Anna, and Mary's firstborn son, Yeshua (4 BCE–72 CE).[9] Joseph was intelligent and eager to learn. He learned various languages with the greatest of ease; in addition to his native Aramaic, he learned Hebrew, Persian, and Sanskrit at a young age. This was followed by Greek and Egyptian and finally, Anna reluctantly taught her son Latin when he was eleven years old, because after much bloodshed, in 63 BCE the Romans occupied Palestine and incorporated the country into their province of Syria.

Anna described how from the age of twelve, Joseph studied in the great library of Alexandria and underwent initiations in the temples on

---

9. Heartsong, *Anna, the Voice of the Magdalenes*, app. C, 379.

the Nile. He also traveled to India. He returned to Carmel in his early twenties. He brought the scrolls he took with him from Tibet, India, and Mesopotamia to the Essene communities in Carmel and Qumran. But both Carmel and Qumran became unsafe; Qumran was burnt to the ground in 37 BCE, and in 31 BCE they had to deal with a major earthquake. Joseph planned to dismantle the libraries in Palestine and transferred scrolls to Essene communities based in mountainous communities around the Mediterranean and the British Isles.[10]

Through Claire Heartsong, Anna described that in the year 32 BCE Joseph, now an adult, left for England at the invitation of the Druid Council. This was followed by annual trips to Ireland, Scotland, and the northern Mediterranean coast. Immediately after returning to Palestine from his first pilgrimage to England, Joseph conceived a plan to organize a fleet of freighters to transport tin and lead. He was the co-owner of two mines in England and was now the Minister of Mines in the Roman Empire. This trade provided a perfect cover for secretly transporting documents to Essene libraries in England and other libraries in the mountainous areas around the Mediterranean. Joseph became an entrepreneur and developed his diplomatic abilities. Ten years later, with the help of the Druidic Councils, he owned a fleet of twelve ships and made annual trips to England.[11]

Anna explained that Joseph became wealthy and had influence within the administrative structure of the Roman Empire. He attracted the attention of a Hasmonean prince with the name of Arimathea; this prince was a member of the highest Jewish legislative body, the Sanhedrin, and had inherited a large estate in Samaria on the road between Galilee and Judea. He became so fond of Joseph that he asked him to marry his daughter, Eunice Salome. The marriage was solemnized in Jerusalem in the year 29 BCE. Arimathea taught Joseph to mediate between warring factions. He also adopted his son-in-law as an adoptive

10. Heartsong, *Anna, Grandmother of Jesus*, 56–57.
11. Heartsong, *Anna, Grandmother of Jesus*, 57–58. In Hebrew, the island of Albion was called *Brith-ain* or The Covenant Land.

son. After this, Joseph became a representative of Galilee and Samaria in the Sanhedrin in Jerusalem. Soon after, he moved to Jerusalem to mingle with the wealthy and literate. There he served for many years as an influential advisor. He formed the link between many groups and sects that were often at odds with each other. His knowledge of languages and his insight into human nature meant that he was respected by everyone. Though cool and reserved at times, he was always there for people in need and for those members of the Essene brotherhood and sisterhood who knew him as an initiate. In Jerusalem he owned three residences that provided him access to the underground passages, known to the secret orders of the brothers of Solomon. He made himself useful to large groups of people in the outside world, but he was also devoted to the hidden orders of the brotherhoods and sisterhoods of Light.[12]

Twelve years after the death of his first wife, Joseph acquired an estate near Magdala on the west coast of the Sea of Galilee. Here, he met a woman of high birth, named Mary. As a young teenager, she had rebelled against the strict discipline of her high-priest father and had married a Macedonian mercenary soldier named Philippus. With him she had three children, Thomas, Matteus, and Susanna. Her husband was not home much, and when he was home he insulted her. Several years after Philippus left Mary and her children impoverished and penniless, Mary of Magdala met Joseph of Arimathea in the marketplace. He asked her to become the head of household in his new home in Magdala.[13] One thing led to another, and in 5 BCE they were married; in the year 4 BCE Mary Magdalene was born. In addition to their eldest daughter Mary Magdalene, Joseph and Mary also had a son, Lazarus, born in 1 BCE, and another daughter, Martha, who was six and a half years younger than Mary.

In Carmel, young girls received preliminary training in the light conception, also known as light birth.[14] There were also couples in

---

12. Heartsong, *Anna, Grandmother of Jesus*, 58–59.
13. Heartsong, *Anna, Grandmother of Jesus*, 86.
14. Heartsong, *Anna, Grandmother of Jesus*, 101.

Carmel who prepared together for a light conception. This phenomenon has recently been scientifically investigated by Marguerite Rigoglioso. I will come back to it in more detail later (see An Esoteric Interlude).

According to Anna, the people who trained for a light conception included:

- Mary Anna and Joseph, who begat Yeshua in a light conception.
- Joseph of Arimathea and Mary of Magdala, who, through a light birth, begat Mary Magdalene, the twin soul of Yeshua.
- Anna's son Isaac with Tabitha, who begat Sarah.
- Anna's daughter Rebekah with Simeon, who begat Mariam, a cousin of Yeshua. When Rebekah died, Mary Anna took her sister Rebekah's daughter, Mariam, into her family and adopted her, so while Mariam was a cousin of Yeshua, she became more like a sister. The two got along together exceedingly well and enjoyed a pure spiritual relationship.[15]
- The couple Elizabeth and Zacharias did not belong to this Carmel group, but Elizabeth did receive her training in Carmel, after which she returned to Bethlehem so she and her husband begat John through a light conception. They were also related to Mary Anna's husband Joseph through his brother Jacob. Jacob's eldest daughter Elizabeth married Zacharias and Jacob's youngest son Joseph was married to Mary Anna.[16]

The Regression Reports

Joanna Prentis questioned in regression both Daniel Benezra and the person who remembered being Joseph of Arimathea about the situation surrounding Joseph of Arimathea. From their information it appeared that Joseph of Arimathea was secretly an Essene; he was a member of the core group who held the deepest of secrets. Apart from the few leaders of the

15. Heartsong, *Anna, Grandmother of Jesus*, 107.
16. Heartsong, *Anna, Grandmother of Jesus*, 98.

main and subcommunities within the Essene tree (which I'll discuss in more detail later in this book), not a single Essene knew of the existence of the core group and his membership in it. Joseph kept all this completely hidden from the outside world. Even during regression, he had difficulty expressing himself, although he was now stripped of his obligation of secrecy. Operating in covert ways to protect his family was deep in his genes.[17] When Prentis wanted to know more about the fate of Joseph after his flight from Palestine (discussed in chapter 14), she consulted Alariel, which added to the information from Daniel and Joseph. I summarize it below but use Alariel's own words as much as possible.

Alariel said:

The commercial merchant fleet of Joseph of Arimathea contained between 120 and 150 ships. Because a number of ships were constantly undergoing maintenance or repair, this meant that the total fleet was never fully operational. Joseph built in surplus capacity in his fleet for two reasons. First, he could respond quickly to urgent orders from the Roman army. Second, he kept ships to transport essential passengers and students when necessary. He never took too heavy a load and therefore did not cause the crew extra stress; he maintained wide margins and as a result, shipping via his ships ran smoothly. This enabled him to recruit the best people; they were loyal to him more than to any other shipowner and he treated his people fairly.[18]

Commenting on this information from Alariel, Stuart Wilson wrote that Joseph's merit lay not only in the efficient mining of tin but also in its reliable delivery to ports within the Roman Empire. He controlled most of the tin mines in Cornwall and had a dominant position in supplying tin to the Romans. Reliable tin supplies were of great military and strategic importance to the Romans, so let's check the historical information: Is this alternative information supported by historical evidence?

---

17. Wilson and Prentis, *The Essenes*, 82–83.
18. Wilson and Prentis, *The Magdalene Version*, 44–45.

A TIN INTERLUDE:
THE TIN TRADE IN ENGLAND

Using historical sources, various English researchers had completed intensive research into the ties of Joseph of Arimathea with England and in particular with the tin mines in southwest England. Among them were John William Taylor,[19] George Jowett,[20] Cyril C. Dobson,[21] Lionel Smithett Lewis,[22] E. Raymond Capt, Robert Mock, and Isabel Hill Elder. In my opinion, their research places the large number of English

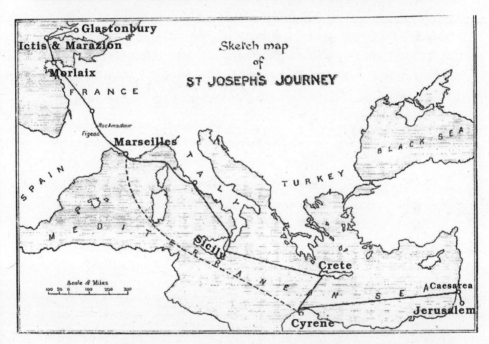

Figure 4.1. Overview of the various sea routes and the route through France taken by Joseph from Arimathea to southwest England. From John Taylor, *The Coming of the Saints*, 1907, 224–225.

---

19. Taylor, *The Coming of the Saints*, 178.
20. Jowett, *The Drama of the Lost Disciples*, 17, 41–42.
21. Dobson, *Did Our Lord Visit Britain as They Say in Cornwall and Somerset*, 13–20.
22. Lewis, *St. Joseph of Arimathea at Glastonbury*, 31–32.

legends that survived in England to this day around Joseph of Arimathea, Mary, Jesus, and Mary Magdalene in a credible context. Joseph was said to have taken the young Jesus with him to England on his travels. He was also said to have settled in England after the crucifixion and his flight from Palestine.

There were three possible transport routes for Cornish tin. One route is entirely by sea and runs through southern Spain to ports in the Mediterranean. Another route runs via the Bay of Biscay to the mouth of the Loire and south via this river. Then it goes overland to Gadiz (Cádiz) in southern Spain and from there by ship to ports in the Mediterranean. The last route goes overland from Morlaix to Marseille and then by ship to Ostia and/or other Roman ports around the Mediterranean. It is believed that this final route was the shortest, despite the fact that the tin had to be transported across France via pack horses.

Starting in 2000 BCE there was a lively tin trade between Phoenicia and Cornwall (See fig. 4.1 for a map of the sea route through Europe and the ports of southwest England). The tin trade was frequently mentioned by classical writers such as Diodorus Siculus and Julius Caesar (see A Second Tin Interlude). In 445 BCE the Greek author Herodotus called Britain the Tin Island or the *Cassiterides*. England was the main source of tin for the Romans.

In the Latin translation of the canonical Gospels of Mark and Luke we find Joseph of Arimathea referred to as *Decurion*; this was a title used by the Romans to designate someone charged with the mining of metals. In the Latin translation of the Bible by Hieronymus (342–420), Joseph's official title was that of *Nobilis Decurio*. This indicates that he held a prominent position within the Roman administration as Minister of Mines. George Jowett wrote that this was not surprising given Joseph's remarkable character traits:

We know that he was an influential member of the Sanhedrin or Jewish council, the religious body of the Jews in Roman times. And that he was a legislative member of a provincial Roman senate. His financial and social status can be deduced from the fact that he owned a palace in the holy city and a magnificent country residence outside Jerusalem. Further north, he owned another vast estate in Arimathea, known today as Ramalleh. It was on the busy caravan route between Nazareth and Jerusalem. All that is known of him characterized him as a wealthy and influential man within both the Jewish and Roman social hierarchies.[23]

23. Jowett, *The Drama of the Lost Disciples*, 17–18.

5
MARY ANNA,
MOTHER OF YESHUA

Little is known about Mary, the mother of Yeshua, from only the biblical sources.[1] Anna of Carmel called her daughter Mary Anna, so that is the name used in this chapter.

According to biblical sources, Mary Anna was Joseph's betrothed, and Isaiah predicted she would have a child.[2] During her early pregnancy she visited her cousin Elizabeth and sang the Magnificat;[3] it follows the well-known Christmas story.[4] According to Jewish custom, Jesus was circumcised in the Temple in Jerusalem eight days after his birth. Simeon predicted that a sword would pierce Mary's heart.[5] Matthew recorded the visit of the wise astrologers and the flight to and return from Egypt.[6] After that there was a gap of twelve years, until Mary returned to the temple in Jerusalem, where she found her twelve-year-old son after a long search.[7] Then, there is another gap of many years. The next time we see her, she attended the famous marriage

---

1. Freedman, Myers, and Beck, *Eerdmans Dictionary of the Bible*, "Mary," 863–65; 863 gives a list of uncertainties and contradictions.
2. Mt. 1:18–25; Lk. 1:26–28; Is. 7:14: "Behold the virgin shall conceive and bear a son and call his name Immanuel."
3. Lk. 1:39–56.
4. Lk. 2:1–20.
5. Lk. 2:33–35.
6. Mt. 2:11, 13–15, 19–23.
7. Lk. 2:41–52.

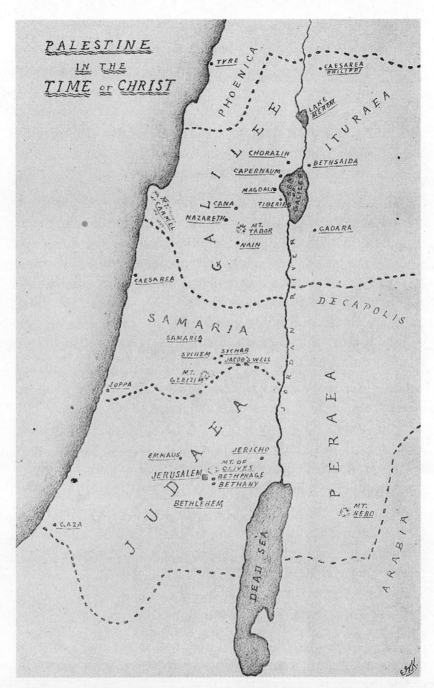

Figure 5.1. Map of Palestine showing Mount Carmel and Nazareth.
From Rev. E. G. Krampe, *Bible Manual: Introductory Course on the Bible*,
Cleveland, OH: Central Publishing House, 1922, 210.

in Cana.[8] Thereafter, Mother Mary appeared a number of times during her son's public life,[9] including standing under the cross at Jesus's crucifixion.[10] During Pentecost, Mary was present in the original church in Jerusalem.[11]

In the apocryphal source *The Gospel of The Beloved Companion*, it is said that the wedding at Cana is the wedding of Yeshua and Miryam of Bethany.[12] In another important episode, Mary Anna stood under the cross, and in *The Gospel of The Beloved Companion*, Mary Magdalene was with her. Jesus linked the two by asking his mother to adopt his wife as a child and asking his wife to take his mother into her home.[13] They both played a role in washing and caring for Yeshua's body before he was placed in the tomb.[14] Finally, Mary Anna was present in Bethany when Mary Magdalene returned with the good news that she had met Yeshua at the open tomb.[15]

There is an apocryphal source that gave a lot of information about Mary (and is also discussed in the chapter about Anna): The Birth of Mary. This text was later given the misleading name The Protevangelium of James or The Infancy Gospel of James, circa 150 CE. The text strongly emphasized that Mary was a virgin when she gave birth.

When Mary was three, the girl was taken to the temple, a detail that was reflected in the channeled information as well, but there it was not the temple in Jerusalem, but the Essene Temple on Mount Carmel. As described in The Birth of Mary, on the third step of the altar Mary danced with joy; in the temple of Jerusalem she was like a dove being fed from the hand of an angel.

When she was twelve years old, she, according to Jewish religious

8. Jn. 2:1–5.
9. Jn. 2:12; Mk. 3:20–21; Lk. 11:27.
10. Jn. 19:25–27.
11. Acts 1:14.
12. *GBC* [6:3, 7, 9], 18–19; *MMU*, 113–125.
13. *GBC* [39:3], 73; *MMU*, 436–38.
14. *GBC* [40:2], 74; *MMU*, 458–59.
15. *GBC* [40:7], 75; *MMU*, 465.

law, was not allowed to "defile" the temple once she started her period. Mary was entrusted to the care of the widower Joseph. The angel visited her twice announcing her upcoming pregnancy. In her third month she visited her cousin Elizabeth for three months. Mary grew bigger and hid because she was ashamed of her pregnancy. The priests believed that Joseph and Mary, as unmarried people, had intercourse. Joseph had a dream in which the angel told him that the child was of the Holy Spirit. Both must now drink the bitter test water, that could cause great pain and distortion of the lower body, and both remain healthy. This was proof that they spoke the truth.

The pair made their way to Bethlehem and found a cave where Mary gave birth with the help of a Hebrew midwife. Everything came to a standstill: time, the stars, and even the birds hung still in the sky. The midwife came across a woman named Salome and told her a virgin had given birth to a child. Salome didn't want to believe this, so she examined Mary and experimentally established that Mary was indeed a virgin. Then Salome's hand fell off, but she touched Jesus and was healed.

Traditionally the idea of the virgin conception and the virgin birth were strongly emphasized. In terms of virgin conception, Mary and Jesus were conceived by the Holy Spirit. In virgin birth, the hymen remains intact even after birth, something that seems completely unnatural. What is up with the cult of virginity? Can the alternative sources provide more clarity on this?

The Channeled Sources

Anna, Mary's mother, provided the following channeled information through Claire Heartsong. When Mary Anna was born in Ephesus after a painless delivery, Anna considered her the Mother of the New Covenant. Anna already knew that in addition to Mary Anna and her partner, more women and men would come to Carmel later to be instructed about light conception together. The children who would be born through light conception would all—of their own free will and choice—play an important part in the events to come.[16]

Mary Anna was born in September of the year 20 BCE in Ephesus. Mary Anna stayed in Ephesus until the age of three (17 BCE). During that time, Mary Anna learned Aramaic, Hebrew, and Greek.[17] In Ephesus, Joachim was engaged in supporting Joseph of Arimathea, who visited the family several times a year. He escorted Essene brothers with their manuscripts from Carmel and Qumran, finding new hiding places on the rugged Aegean islands. Then he took them back to Galilee and Judea.[18]

Shortly before Mary Anna's third birthday in 17 BCE, the family returned to Carmel. There she was consecrated to the Lord Most High on her third birthday. Her family handed her over for training and education at the Carmel mystery school. People from Tibet, India, Mesopotamia, Egypt, Greece, and England came there to instruct her. Anna was in charge of the young girls' school and saw her daughter every day.[19]

Mary Anna started menstruating at the age of twelve. Anna took her along with other young girls to the cave dedicated to the Great Mother. She initiated the young girls, explaining the blood mysteries and the opening of the womb. There was also a stone circle. The girls listened to stories, meditated, and learned rituals, including how to channel the energies of the divine feminine. They were prepared for marriage and—in the literal words of Anna—tantric love.

Between the ages of twelve and fifteen, Mary Anna was regularly taken to the temple in Jerusalem, where she trained for three months. In this way, she was prepared for the customs of the larger Jewish community, which differed greatly from those of the Essene community.

Both Anna and Joachim saw in a vision that the time of the light conception was approaching for the group of young girls and women in training. Along with Mary Anna there were fourteen girls and nine women in training that season. Among the men, there were nineteen

---

16. Heartsong, *Anna, Grandmother of Jesus*, 89.
17. Heartsong, *Anna, Grandmother of Jesus*, 91.
18. Heartsong, *Anna, Grandmother of Jesus*, 92.
19. Heartsong, *Anna, Grandmother of Jesus*, 92.

boys and five men who underwent an initiation on Carmel. Five of them were also willing to fulfill their partner role in light conception: both the male and female candidates were trained to hold high frequencies of light in their consciousness and mind. They learnt how to discern energies, how to astral travel, how to bilocate, and how to focus their attention on certain dimensional focus points for days on end.[20]

Anna's channeling explained that one of the men was Joseph, son of Jacob (37 BCE–20 CE). He was born in Bethlehem as the youngest son of Joachim's brother Jacob. He was thirty-two years old and a widower for five years. He had undergone advanced initiations in Egypt and in India. He was a true son of Zadok according to the order of Melchizedek, who was a creature that sometimes appeared out of nowhere. Joseph was initiated by this mysterious person and he carried the priestly rod of Zadok. It was carved from a branch of the hawthorn and had precious stones laid into it and symbols carved into it.[21] In their visions, Anna and Joachim saw that the Anointed One would be born through Mary Anna and Joseph.

During a solemn ceremony, both Mary Anna and Joseph made it clear that they had chosen each other. They were blessed by the community to be engaged, and they married at the end of the year 5 BCE. Other couples also stepped forward. Among them were Joseph of Arimathea and his beloved Mary of Magdala, whom he had met three years before.[22]

In the last week of June of the year 5 BCE Anna and Joachim's room was flooded with an immense light from four angels. They were Michael, Gabriel, Raphael, and Uriel. Anna and Joachim were lifted into a light chamber they recognized from when they conceived their children. Other couples also entered this light room, including Mary Anna

---

20. Heartsong, *Anna, Grandmother of Jesus*, 102.

21. Heartsong, *Anna, Grandmother of Jesus*, 104.

22. Heartsong, *Anna, Grandmother of Jesus*, 103. In 29 BCE Joseph married Eunice Salome, with whom he had two daughters. His first wife died in 20 BCE, and then he married Mary of Magdala in 5 BCE. Their daughter, Mary Magdalene, is born in the year 4 BCE.

and Joseph, and Joseph of Arimathea and Mary of Magdala. They were told that these couples would participate in light conception. But like Anna with Joachim, Mary Anna did not need the physical DNA of Joseph in her light conception.[23]

Anna went on to revoke the Gospel story, saying that Joseph would have been ignorant of what had happened to Mary Anna, and that she would have known little of the conception of light. The records that stated the truth had been hidden. They were suppressed by the church and state and largely destroyed, but according to Anna it will be known again in a new period of light.[24] I will return to the phenomenon of light conception in more detail in chapter 7 when discussing Mary Magdalene (An Esoteric Interlude).

The formal wedding feast of Mary Anna and Joseph was celebrated in the late autumn of the year 5 BCE.[25]After this, Mary's father died. Yeshua was born in Bethlehem according to the Jewish calendar on Nisan 21, one hour after midnight. According to the Gregorian calendars, this was early in the month of April of the year 4 BCE. That means Mary Magdalene was born in a remarkable synchronicity that same day at 11:00 p.m. There will be more information about this later.[26]

Anna explained that Mary Anna and Joseph had seven children: Yeshua, born in Bethlehem in 4 BCE; James and Jude, born in Heliopolis in 2 BCE; Joseph the younger, born in Heliopolis in the year 1 BCE; Ruth, born on Mount Carmel in the year 4 CE; Thomas and Simon, born in Nazareth in the year 7 CE. Finally, they adopted Mariam, granddaughter of Anna through her daughter Rebekah (Mary Anna's sister). Joseph died in 20 CE and made his ascension in the Himalayas, accompanying Yeshua on his journey to India between 14 and 21 CE.[27]

Later in Nazareth, Mary Anna married Ahmed, an Egyptian

23. Heartsong, *Anna, Grandmother of Jesus*, 105.
24. Heartsong, *Anna, Grandmother of Jesus*, 107.
25. Heartsong, *Anna, Grandmother of Jesus*, 109.
26. Heartsong, *Anna, Grandmother of Jesus*, 119, 121.
27. Heartsong, *Anna, Grandmother of Jesus*, 286.

Essene who was born in Heliopolis in 23 CE. They moved to Jerusalem in the year 27 and had three children: John Mark, born in Nazareth in 24 CE, and the twins, Esther Salome and Matteas, born in Nazareth in the year 25. Ahmed died in Jerusalem in the year 31. Mary Anna fled in 32 CE and arrived in the south of France that summer in Saintes-Maries-de-le-Mer. From there they traveled in early October to the community at Mount Bugarach. She made numerous journeys from here and consciously died at the age of eighty-five near Ephesus, the city where she saw the light of day, in the year 66.[28]

The Regression Reports

The information about Mary, the mother of Yeshua, from the regression accounts is discussed in chapter 8 of this book, when the family was still in Palestine and not yet in France. In short, according to the regression reports, Mary Anna traveled to Egypt to supervise the priestesses in training. She was part of the ultra-secret core group that guided the mission of Yeshua and Mary Magdalene and was the head of the first circle of female disciples. Mary Anna and Mary Magdalene differed in character, but both were fully committed to their common mission.

28. Heartsong and Clemett, *Anna, the Voice of the Magdalenes*, 368, 373.

6
YESHUA

It is curious that the canonical Gospels—after the mention of the birth of Yeshua and his presentation as a newborn in the temple in Jerusalem—do not give any information again until Yeshua went with his parents to the temple in Jerusalem when he was twelve years old.[1] There is a gap of twelve years, except for a single general remark in Luke.[2] Then another long period is skipped and we do not meet the biblical Yeshua until he is thirty, as he entered the public years[3] and began preaching after being baptized by John.[4] That leaves a gap of eighteen years. Apart from these short fragments, information for about twenty-nine years of Yeshua's life is missing. The big question is what happens in the lost years?

There are some apocryphal texts that focus specifically on Yeshua's childhood and emphasize the miracles he performed.

- The story of Mary's birth and that of her firstborn son Yeshua can be found in The Birth of Mary from around 150 CE; this was later renamed The Protevangelium of James or The Infancy Gospel of James.
- The Infancy Gospel of Thomas from the second century described

---

1. Lk. 2:41–52.
2. Lk. 2:40: "And the child grew and became strong, filled with wisdom; and the favor of God was upon him." See also Lk. 2:52: "And Jesus increased in wisdom and in stature, and in favor with God and men."
3. Lk. 3:23.
4. Mt. 3:13; Mk. 1:9; Lk. 3:21; *GBC* [4:1], 15.

Yeshua's childhood up to his appearance as a twelve-year-old in the temple. It contains the famous story about the clay birds that flew up on the Sabbath when the five-year-old child Jesus clapped his hands.[5]

• The Arabic Infancy Gospel of Jesus dates from the fifth century; it paid a remarkable amount of attention to Jesus's stay in Egypt.[6]

• In addition, The Pseudo-Gospel of Matthew from well after the fifth century also gave experiences from Yeshua's stay in Egypt.[7]

The Channeled Information

Anna, Yeshua's grandmother, provided the following information: On the eighth day after his birth in the year 4 BCE in Bethlehem, Yeshua was circumcised and registered in the synagogue in Bethlehem. Six weeks after his birth, he was taken to the temple in Jerusalem. Two turtledoves were sacrificed, despite the fact that the Essenes refrained from animal sacrifices.[8] An elder named Simon recognized Yeshua. When the aged High Priestess Anna stepped forward to hold the child in her arms, she recognized Mary Anna from her previous visits to the temple in Jerusalem (when Mary Anna was between the ages of 12 and 15); this woman had been a substitute mother and teacher to Mary and understood who she held in her arms because she had foreseen everything, but she agreed she kept her mouth shut.[9]

Anna wrote that the family met in Joseph of Arimathea's beautiful house near the temple. He announced that wise men had gathered with Herod in Jerusalem along with their wives. Joseph received them

5. Hennecke and Schneemelcher, eds., *Neutestamentliche Apocryphen*, vol 1, 1964, 353; *Het Grote Boek der Apocriefen*, Deventer, The Netherlands: Ankh-Hermes, 2009, 188–95, 189.
6. Hennecke and Schneemelcher, *Neutestamentliche Apocryphen*, 365; *Het Grote Boek der Apocriefen*, 195–204 with fragment about Jesus in Egypt.
7. Hennecke and Schneemelcher, *Neutestamentliche Apocryphen*, 367; *Het Grote Boek der Apocriefen*, 176.
8. Heartsong, *Anna, Grandmother of Jesus*, 123.
9. Heartsong, *Anna, Grandmother of Jesus*, 124.

but kept the family's presence a secret. If danger arose, the little family would leave through a secret underground passage that led to the Garden of Olives outside the city wall. The reception room in the upper room had been used by the Essenes for centuries; in danger they can leave the city entirely through tunnels under the floor.[10] While waiting for the magi, a small family reunion was organized that was also attended by Mary of Magdala of Bethany, her sister-in-law Martha, and her young baby daughter Mary of Bethany or the later Mary Magdalene. After three days, the magi arrived and solemnly entered the reception room. They bore gifts, and instead of going back to Herod, they decided to leave the city.[11]

The angel Gabriel appeared in a dream to Joseph, son of Jacob, and warned him not to return to Carmel but to go straight to Egypt and wait there for a sign of the return. That very night, Joseph of Arimathea summoned a trusted friend and ordered messages to be sent to all the families of the brotherhood and sisterhood to save themselves. Elizabeth was warned to take John from Bethlehem to Martha's safe house in Bethany. A ship was waiting for Mary Anna and her family to travel to Alexandria. There they were taken in by members of the brotherhood and sisterhood of Light.[12]

They were secretly taken to Heliopolis via Alexandria. Just before Yeshua's first birthday, Anna's son Isaac and his wealthy wife Tabitha gave them accommodation in their house nearby. Yeshua resided there for the next seven years of his life, surrounded by a crowd of loving aunts and uncles. In the summer of the year 2 BCE Mary Anna gave birth to twins, James and Judas. The following year, Joseph the younger was born. Also present were Sara, the daughter of Isaac and Tabitha, and Mariam, the daughter of Rebekah and Simeon, with whom Yeshua had a good relationship.[13] When Mariam's mother Rebekah died, Mary Anna adopted the girl and Mariam became Yeshua's sister.

---

10. Heartsong, *Anna, Grandmother of Jesus*, 124–25.
11. Heartsong, *Anna, Grandmother of Jesus*, 126, 128.
12. Heartsong, *Anna, Grandmother of Jesus*, 128.
13. Heartsong, *Anna, Grandmother of Jesus*, 128–29, 133.

In Egypt, Yeshua learned all kinds of languages at a young age. He was fluent in Egyptian and Greek, as well as his family languages of Aramaic and Hebrew. He liked to research ancient papyrus texts and went with his father to the libraries, pyramids, and temples along the Nile. Sometimes Mary Anna went along and underwent initiations in the temples of Isis, Hathor, and Horus. She also underwent an initiation into the Great Pyramid. Yeshua taught from childhood about a new way of personal and planetary ascension, which he later demonstrated publicly. He was clairvoyant and had a gift for healing. By his seventh year, his body was tall and slender. Next to the house that Joseph and Simeon rented from the wealthy father of their sister-in-law Tabitha, was a cave where Yeshua often retreated.[14]

In February of the year 5 CE, Joseph received the message from the angel Gabriel that it was time to return—King Herod had died. Weeks before Yeshua's eighth birthday, Mary Anna and Joseph held a birthday party, which was also a farewell party. On their way back they pilgrimaged to Mount Sinai. The group leaving Heliopolis consisted of Joseph, Mary Anna (who was pregnant with her fifth child), and their children Yeshua, James, Judas, and Joseph the younger. In addition, Simeon and Rebekah left with their daughter Mariam and numerous other adult cousins. There were helpers accompanying the caravan of camels, donkeys, and ox carts.[15] Via the Sinai they followed the old route to Hebron, and from there via Bethlehem to Mount Carmel. Yeshua resided in Egypt from the year 4 BCE until the year 4 CE.[16]

Yeshua moved between Carmel, Nazareth, and Qumran (see fig. 5.1, map) but he lived on Mount Carmel between the year 4 and the year 6 CE, then from the year 6 to the year 9 in Nazareth. From both places he traveled to Qumran. When he was in Carmel, he spent much of his time with his grandmother, Anna. He underwent numerous initiations in Carmel, but also in Qumran. On Mount Carmel, he

---

14. Heartsong, *Anna, Grandmother of Jesus*, 129, 130, 132.
15. Heartsong, *Anna, Grandmother of Jesus*, 133.
16. Heartsong, *Anna, Grandmother of Jesus*, 288.

lived with the older boys and celibate men in a communal dormitory. He was treated like everyone else. He spent long hours in the library, reading, translating, and copying ancient texts from Alexandria, Greece, Persia, India, and the Himalayas. He learned the different languages quickly and liked to enter into a dialogue about the content with people who came in. Long before the sun rose, he walked in nature or sat in meditation under the cedar trees or was engrossed in a text. Like his father Joseph, he slept little. He did all his assigned chores, such as cleaning the floors, latrines, and the kitchen without complaining, and also with great sensitivity, gentleness, and a light heart. In Egypt, he learnt to make music, and by the age of twelve he was an accomplished singer, songwriter, and instrumentalist.[17]

During the latter part of the year 5 CE, Joseph settled in a new village called Nazareth, named after the Nazarites (a subdivision of the Essenes). It was a short day's walk from Mount Carmel and near Cana in Galilee where Joseph's parents lived. Nathan, Mary Anna's brother, owned land here. He suggested that Joseph built a house on this spot. At the beginning of the year 6 CE they moved there. Joseph was a master carpenter in fine woods and made musical instruments and tabernacles. He often took Yeshua along as an assistant.[18] Anna said that while in Carmel, Yeshua wore his hair like a Nazarite: long and parted in the middle.

Yeshua was consistently referred to as Jesus the Nazorean in the Gospel of Miryam, while the canonical texts mostly speak of Jesus of Nazareth. In doing so, they obscure the basic meaning. The name Nazorean refers to the Hebrew word *neser* or shoot or scion of Jesse, the father of King David. The verb *nsr* also means "to guard" and "to protect." It is therefore about the "keepers of the covenant."[19] From Anna's remark it can be deduced that a community of Nazarites founded the new village of Nazareth and that this was the basic reason for the family to settle there.

---

17. Heartsong, *Anna, Grandmother of Jesus*, 144. He sings the psalms of Akhenaten, David, and Zarathustra. He performs the dialogue between Krishna and Arjuna.
18. Heartsong, *Anna, Grandmother of Jesus*, 144.
19. *MMU*, 113nn7–11.

Yeshua underwent his second rite of passage to adulthood at the age of twelve within the new synagogue in Nazareth, which Joseph helped build. He amazed the leaders with his knowledge and wisdom about the Torah and other ancient texts such as Enoch, Zadok, and Moses. After this there was another gap of twelve years before Yeshua appears before the priests of Levi in Jerusalem. This visit was about the ritual to adulthood, as it was performed by the Jews in those days, according to Anna.[20]

During his twelfth year of life, Yeshua became more deeply involved in the Carmel mystery school. Anna said she taught him to hold his breath and let the breath of life circulate through his body until his heartbeat slowed to an inaudible whisper. "In the caves of Mount Carmel, I taught him how to manage his *prana* and how to move through the veils of the astral world. Some caves have been hollowed out to produce specific acoustic sounds. Deep-trance states are reached when these sounds are made conscious and heard."[21] Anna said she used her voice and brought her frame drum, the sistrum, and bronze bells. Anna said that Yeshua brought water, sleeping supplies, oil lamps, and herbs to support their vigils. Sometimes they stayed in the caves for a week. Together they made ethereal journeys to the inner abodes of the masters. Some of their favorite places to bilocate to were places in England. "I promise him that one day he would physically travel to England and undergo initiation at the hand of his uncle Joseph of Arimathea."[22]

According to Anna, Yeshua spent time in England from 9 CE to 12 CE (see An English Interlude and A Celtic Interlude), and by his thirteenth birthday Yeshua underwent the first level of the "rite of the grave" on Mount Carmel. She said, "All our previous work has been aimed at preparing him for the state called 'death' and returning his soul to his body in a process of resurrection. Therefore, I took Yeshua to a secret room behind a false door, in which I had placed my body from time to

20. Heartsong, *Anna, Grandmother of Jesus*, 144.
21. Heartsong, *Anna, Grandmother of Jesus*, 145.
22. Heartsong, *Anna, Grandmother of Jesus*, 145.

time when I wanted to rest for a longer period of time.[23] I waited long hours and watched all the initiations he would undergo in the future pass before me." Anna described how at the tender age of twelve, Yeshua began his preparation to publicly demonstrate the crucifixion and resurrection initiations. He knew that death was an illusion and with each resurrection initiation that followed in the coming years, he would bring more light into his physical body[24] (chapters 12 and 13).

Anna explained that as he hit puberty and got taller, he also became quieter and more withdrawn. In his meditations he sometimes encountered an energy field of consciousness; he'd asked her to explain these or asked her for further explanation when he discovered contradictions and ambiguity in the scriptures. They often had discussions about the different points of view of the various religious sects, their doctrines, and their practices. She always advised him to go within and look for the answers in his own heart. The period that Yeshua moved between Carmel, Nazareth, and Qumran lasted five years and, as mentioned, ran from the years 4 to 9 CE.[25] But then changes were announced.

Anna explained that his uncle and parents had already agreed that Yeshua would go to England for three years when he was thirteen years old.[26] He stayed there in the years 9 to 12 CE.[27] There he met his uncles Andreas, Josephus, and Noah and his cousin Sara.[28] His other uncle, Joseph of Arimathea, was active on the two islands in Avalon of Glastonbury and the larger island of Anglesey off the coast of Wales. He wanted to found Druidic-Essene universities with libraries of thousands of books, manuscripts, and scrolls, many of which he transported to England himself.[29] In the year 12 CE, the then sixteen-year-old Yeshua returned from England to Carmel.

23. Heartsong, *Anna, Grandmother of Jesus*, 145.
24. Heartsong, *Anna, Grandmother of Jesus*, 145.
25. Heartsong, *Anna, Grandmother of Jesus*, 288.
26. Heartsong, *Anna, Grandmother of Jesus*, 167.
27. Heartsong, *Anna, Grandmother of Jesus*, 288.
28. Heartsong, *Anna, Grandmother of Jesus*, 157.
29. Heartsong, *Anna, Grandmother of Jesus*, 158.

AN ENGLISH INTERLUDE: YOUNG YESHUA IN ENGLAND

The historical sources, such as those by the Reverend Baring-Gould, are at the forefront of research into the legends and historical sources in Cornwall. In the nineteenth century he was interested in the folklore of Cornwall.

The Legends of Cornwall Explored by the Reverend Baring-Gould

He collected what he found and published *The Book of Cornwall* before his death in 1923.[30] In Cornwall, Joseph of Arimathea was known to come to Cornwall in a boat with the infant Jesus and teach him how to extract tin from rock. Baring-Gould recorded that, when the tin caught fire, the tinsmith shouted, "Joseph was in the tin trade."[31]

The Legends of Cornwall Explored by the Reverend H. A. Lewis

Another clergyman followed this marked trail. The Reverend H. A. Lewis was the vicar on St. Martins, one of the Isles of Scilly. He collected traces of the legend that the young Jesus visited Britain, at fourteen different places on the coast and inland in Cornwall. In 1939 he wrote the book *Christ in Cornwall and Glastonbury, the Holy Land of Britain.* He had meticulously researched memories among the local population; this was reported on The Ensign Message website.[32]

---

30. Baring-Gould, *A Book of Cornwall*, 57.

31. Baring-Gould, *A Book of Cornwall*, 57; Taylor, *The Coming of the Saints*, 180–82, 224 (gives several examples of the presence of Hebrew people in Cornwall and how they are connected with the tin trade; Prophet, *Mary Magdalene and the Divine Feminine*, 250.

32. "Christ in Cornwall and Glastonbury, the Holy Land of Britain," available at the Ensign Message website.

The Legend in Somerset Explored by the Reverend Cyril Dobson

Following the above publications, the Reverend Cyril Dobson went to Somerset—the region north of Cornwall, including the town of Glastonbury—to investigate. In 1999, he published *Did Our Lord Visit Britain as They Say in Cornwall and Somerset?* In the Somerset or Summerland region he researched the traditions of that region, including place names, folklore, and the old mining traditions in the Mendip Hills north of Glastonbury. He transmitted the following information.

- There is a tradition from Somerset in which Joseph of Arimathea resided in the Mendip Hills in the mining village of Priddy. There is an old saying, "As sure as our Lord was in Priddy."[33]
- There is a legend in Somerset that Jesus and Joseph came to Somerset on a ship from Tarshish, an ancient port near Cádiz in southern Spain that must also have been part of Joseph's overseas trade network. Tarshish appeared twenty-five times in the Hebrew Bible. There are Bible passages that refer to the ships of Tarshish as the origin of King Solomon's wealth; he is said to have collaborated with King Hiram of Phoenicia.[34]
- According to another Somerset tradition, the young Jesus and his uncle Joseph resided in Paradise, an old name for Glastonbury and the surrounding area.[35] The ancients experienced Avalon as a gateway to the Other World,

33. Prophet, *Mary Magdalene and the Divine Feminine*, 250.
34. Ez. 27:12 and 25: Large quantities of important metals are said to have been exported from Tarshish to Phoenicia and Israel. The information in the Hebrew Bible describes Tarshish as a source of King Solomon's great wealth of metals, especially silver, but also gold, tin, and iron (Ez. 27). The metals were reportedly obtained in collaboration with King Hiram, Phoenician king of Tyre (Is. 23) and fleets of ships from Tarshish.
35. Prophet, *Mary Magdalene and the Divine Feminine*, 250.

hence the name Paradise. To the west of the town there is an area still called Paradise, as well as a Paradise Farm and a Paradise Lane north of the Tor, which is one of four hills on the former island of Avalon.[36]
- Yet another Somerset tradition said that Jesus and Joseph were at Glastonbury.

The Reverend Cyril Dobson summarized the traditions as follows: Jesus did indeed visit Glastonbury. Joseph became wealthy as an importer of tin and the tin trade, which existed between Cornwall and Phoenicia. On one of his journeys, he took Jesus as a boy with him. Later, Jesus came back to England as a young man and stayed in Glastonbury. He erected for himself a small house of willow and mud, the usual and reliable building materials of the time. Later, after the crucifixion, Joseph fled from Palestine. He settled on that very spot in Glastonbury and built a structure of willow branches and mud, and it would become the first Christian church in the world.[37]

Other Authors
Researchers who took the legends and sources from southwest England seriously argued that Jesus was absent from Judea for extended periods of time during the so-called hidden years.[38] He did not reappear in Judea until he was thirty years old.[39]

36. Michell, *New Light on the Ancient Mystery of Glastonbury*, 79.
37. Michell, *New Light on the Ancient Mystery of Glastonbury*, 80, 138; Prophet, *Mary Magdalene and the Divine Feminine*, 250–52.
38. Michell, *New Light on the Ancient Mystery of Glastonbury*, 80 with ref. to Dobson, *Did Our Lord Visit Britain*; Jowett, *The Drama of the Lost Disciples*, 136; Prophet, *Mary Magdalene and the Divine Feminine*, 249–55.
39. Jowett, *The Drama of the Lost Disciples*, 137: mentions Mt. 17:24 where a tax collector asks Peter if Jesus has paid his taxes, indicating that Jesus is a stranger.

- In 1961, based on the work of earlier researchers, George Jowett emphasized that the traditions of Cornwall, Devon, Somerset, Wiltshire, and Wales all told that Joseph of Arimathea visited Britannia in the company of the young Jesus.[40] But legends are not hard evidence.
- In 1990, author John Michell argued that even the skeptical historians in England are convinced of the antiquity and importance of the legend that Jesus visited Britannia and that Joseph of Arimathea started a community here, but that it can never be proven with precise details.[41]

The following was generally emphasized in English literature: Joseph of Arimathea was the apostle of England and the founder of the English church. It led the great poet William Blake to write a famous poem that Jesus's teaching of love and wisdom came about "upon England's mountains green."[42]

All the above indicates that it is quite possible that the young Yeshua traveled with his uncle Joseph from Arimathea to the British Isles and went to India, where the "holy isles in the west" are known.[43] With these activities, the lost years seem to be filled in meaningfully; the experiences feature areas spanning from the far West to central Asia in the East. I'll return again to the topic of England and India in chapters 12 and 13 when I discuss the events around Yeshua and Joseph of Arimathea after the crucifixion.

40. Jowett, *The Drama of the Lost Disciples*, 69.
41. Michell, *New Light on the Ancient Mystery of Glastonbury*, 82.
42. Jowett, *The Drama of the Lost Disciples*, 147, with ref. to the poem "Jerusalem" by William Blake.
43. Jowett, *The Drama of the Lost Disciples*, 137 gives an Indian source, the Vishnu Purana (see chap. 20 in this book) that says that "the holy islands in the west" are known: people know Britain as a seat of religious education.

AN INDIAN INTERLUDE:
YESHUA IN INDIA

The question is: Does evidence exist to corroborate Yeshua's journeys to the East?

Nicholas Notovitch

Starting in the nineteenth century, travelers from India and Tibet had returned and told the West that Jesus traveled in the East and that records of it had been preserved on-site. One of them was the Russian journalist and writer Nicholas Notovitch, who wrote *The Unknown Life of Jesus Christ*, published in 1894. The book was received with great skepticism in the early twentieth century and rejected as fraudulent. It was even claimed that the book could have been written by an American atheist who never left America, though there was a witness who met Notovitch on his return to Kashmir.[44]

The Church, in particular, made every effort to discredit Notovitch and his work. Notovitch went to Rome and reported on his journey to the East in an attempt to gain support to publish a book about it, but help from an unnamed cardinal was not granted. Notovitch stated that there were at least sixty-three documents in the Vatican that described travelers' testimonies about Jesus's travels to the East.[45] These documents were allegedly taken by missionaries who had been in India, China, Egypt, and Arabia.[46]

Notovitch brought a translation of *The Life of Issa* (Issa is another name for Jesus) from Hemis Monastery, the largest and

---

44. Ahmad, *Jesus in Heaven on Earth*, 355n6, who mentions Sir Francis Younghusband, representative of the British Crown at the court of the Maharaja of Kashmir, as a witness.
45. Notovitch, *The Unknown Life of Jesus Christ*, Introduction; Prophet, *The Lost Years of Jesus*, 107.
46. Prophet, *The Lost Years of Jesus*, 105–07; Ahmad, *Jesus in Heaven on Earth*, 357.

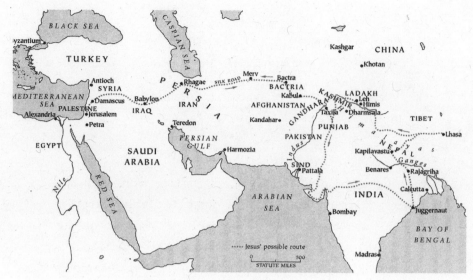

Figure 6.1. Jesus's first journey to India.
From Elizabeth Clare Prophet, *Mary Magdalene and the Divine Feminine*, 32,
courtesy of Summit University Press; Khwaja Nazir Ahmad,
Jesus in Heaven on Earth, 352, gives a similar, more schematic map.

most famous monastery in Ladakh in Northern India bordered
by Tibet, and wrote a book about it. It unleashed a storm of
criticism. I quote the Dutch researcher Paul van Oyen, who
gave the following information about what happened in Hemis
after Notovitch's visit:

> The abbot of the monastery was bothered by all kinds of
> probing questions about the manuscript of Issa. There were
> several attempts to steal manuscripts from the monastery.
> The result of all this nuisance was that for a long time the
> monastery denied that the manuscripts were in the posses-
> sion of the monastery and that Notovitch had been in the
> monastery. This information was, of course, grist to the
> mill of Notovitch's critics; they determined, on the basis of
> the monastery's reactions, that the text must be a creative
> forgery. All this must have sprung from the overactive brain
> of Notovitch, who in their eyes was a charlatan or liar. But

in 1922 all the noise had calmed down . . . and the abbot could speak freely again.[47]

Swami Abhedananda

After all the rejection and fierce criticism from the Church, it took Notovitch almost thirty years to gain acclaim. The first person to support him was Swami Abhedananda, a disciple of the Indian saint Ramakrishna, who spread Vedanta in America. After spending twenty-five years in the United States and having read Notovitch's book, about which he was quite enthusiastic, the swami returned to India in 1921 at the age of fifty-six. He settled in the monastery called Ramakrishna Vedanta Math in Calcutta. He dreamt of making a pilgrimage through the Himalayas on foot. He put his money where his mouth was and wrote a travelogue during the hike, which was edited by an assistant and published in 1929.[48] During his trek through the Himalayas, he decided to visit Hemis Monastery—located forty kilometers north of Leh, the capital of Ladakh, "to verify the story of Notovitch."[49] He wondered if Notovitch really got a translation of an ancient Buddhist manuscript that described Jesus's sojourn in the East as *The Unknown Life of Jesus Christ*. Elizabeth Claire Prophet, who wrote a well-known and valued book entitled *The Lost Years of Jesus*, mentioned what can be read in chapter 13 and 15 of the Hindu swami's account of his visit to the Buddhist monastery.

This chapter in the Bengali book *Kasmir O Tibbate (Kashmir and Tibet)* had to be translated from Bengali into English for Prophet.[50]

47. van Oyen, *De Vergeten Jaren van Jezus in India en Tibet*, 22.
48. Prophet, *The Lost Years of Jesus*, 43 states that the original title was *Parivrajaka Swami Abhedananda*, 1929, which later was republished as *Kashmir O Tibbate (In Kashmir and Tibet)*, 1954.
49. Prophet, *The Lost Years of Jesus*, 42n65, 225.
50. Prophet, *The Lost Years of Jesus*, 44

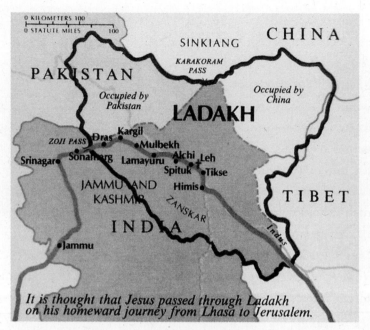

It is thought that Jesus passed through Ladakh on his homeward journey from Lhasa to Jerusalem.

Figure 6.2. Map of Ladakh. From Elizabeth Clare Prophet, *The Lost Years of Jesus*, 342 (1984 ed.), 385 (1987 ed.), courtesy of Summit University Press.

The Swami left at dawn to visit the monastery. He was given an extensive tour, during which he described the monastery in detail in his travelogue. The Swami asked about the manuscript about Issa. The lama who showed the Swami around took a manuscript about Issa from the shelf and showed it to the Swami. He said this was a copy and the original was in a monastery near Marbour in Lhasa; that the original was written in Pali, but that this was a translation into Tibetan.[51]

It consisted of 14 chapters and 224 verses. At the swami's request, a lama helped him translate the text into English. This was later translated into Bengali and had to be translated back to English for Elizabeth Prophet. It is clear that the chemistry

51. Prophet, *The Lost Years of Jesus*, 44n67, 230.

between the chief lama and the swami, an ascetic who had lived for three months in a cave at the source of the Ganges, was different from news-hungry strangers, wrote Elizabeth Prophet. Exactly like Notovitch, the swami did see the text.

Both Notovitch and the swami relied on the same Tibetan document. Both translations—despite all the translations from Tibetan to English, French, Bengali and back to English—barely differ from each other.[52] Paul van Oyen wrote: "This was a great compliment to the precision and care with which Notovitch and the translators had worked."[53]

After this, another disciple of Ramakrishna visited Hemis Monastery. He was told by the lamas that Notovitch had spent some time in the monastery and was given access to the manuscript about Issa. Paul van Oyen also listed eight major Indian sources that supported Swami Abhedananda and confirmed that Jesus resided in India, Nepal, and Tibet. Among them is Paramahamsa Yogananda, known for the famous book *Autobiography of a Yogi*. Among those eight, there is also a confirmation from the early twentieth century from the Abbot of Hemis Monastery, where Notovitch saw the document and had it translated for him.[54]

Nicholas Roerich

Three years later, in 1925, Nicholas Roerich followed in the footsteps of the swami. He briefly visited the Hemis Monastery and was shown an identical text to *The Life of Issa*. Roerich traveled through Central Asia for more than five years between 1924 and 1928. He recorded in his books the oral tradition of Jesus's travels in the East; he received information about this several times during his tour. He also heard

---

52. Prophet, *The Lost Years of Jesus*, 231n1.
53. van Oyen, *De Vergeten Jaren van Jezus in India en Tibet*, 24.
54. van Oyen, *De Vergeten Jaren van Jezus in India en Tibet*, 5–6.

parts of the Tibetan translation of the *Gospel of Issa*, wrote it down, and published it. Legends and parts of *The Life of Issa* as noted by Roerich in his books can also be found *in The Lost Years of Jesus* by Elizabeth Clare Prophet.[55] While travelling, Roerich made about five hundred paintings about the Himalayas in the most beautiful colors. There is a striking similarity between the three translations, including: 1. Notovitch's first translation; 2. the translation of Swami Abhedananda; and 3. the translation of the sixty verses from the *Gospel of Issa* that Roerich mentioned in his book *Altai-Himalaya*.[56] Paul van Oyen wrote: "All things considered, there was sufficient and independent information available to confirm the existence of the *Gospel of Issa*."[57]

Elisabeth Caspari

In 1939, the Swiss musician Elisabeth Caspari also visited the Hemis monastery. She followed in the footsteps of the swami and Roerich, who were there in 1922 and 1925, respectively. The librarian showed her three manuscripts wrapped in colorful cases, which were ceremoniously unwrapped and presented with the words: "These books say your Jesus was here."[58] Therefore, there is every reason to believe that Notovitch—like others after him—made a translation of an authentic Tibetan text that resided in the Hemis monastery. We can therefore take Notovitch seriously.

---

55. Prophet, *The Lost Years of Jesus*, 241–280; see chap. 4 "Legends of the East" by Nicholas Roerich with original excerpts from *The Life of Issa* from Roerich's books, *Altai-Himalaya: a Travel Diary* and *Heart of Asia*.

56. Prophet, *The Lost Years of Jesus*, 53, 229; van Oyen, *De Vergeten Jaren van Jezus in India en Tibet*, 25.

57. van Oyen, *De Vergeten Jaren van Jezus in India en Tibet*, 25.

58. Prophet, *The Lost Years of Jesus*, 58–60, 283–323; Prophet, *Mary Magdalene and the Divine Feminine*, 24.

Notovitch in the Hemis Monastery

In 1887, Notovitch made a tour through Kashmir and Ladakh, which are on the border of India and Tibet (fig. 6.2). In the introduction to his book, Notovitch wrote about how he had a conversation with the abbot of the Buddhist monastery, Mulbekh in Ladakh, which is situated on top of a mountain and built high against a rock face.[59] The abbot told him that in the capital of Tibet, Lhasa, there were thousands of ancient scrolls about the life of the prophet Issa, the Eastern name for Jesus, and that some important monasteries had copies.[60] Based on this information, Notovitch decided to travel to Lhasa to investigate further. He was determined to find records of *The Life of Issa*, even if he had to go to Lhasa to do so. He visited several monasteries, but the monks there said they had no copies. On his way to Tibet, he reached the city of Leh, the capital of Ladakh, and afterward he reached the Hemis monastery. It was the most important monastery in Ladakh; it was located in a remote valley in the mountains, at an altitude of about 3,352 meters. The monastery was extremely difficult to reach and therefore it was spared destruction by Asian conquerors. Numerous other monasteries brought their treasures to this remote place for protection. That was why this monastery, more than other monasteries, had important books, masks, and objects. There was a special secure room called The Dark Treasury that only opened when the departing custodian handed over his position to his successor.[61]

Notovitch witnessed the performance of one of the many mystery plays performed by the lamas. He asked the head lama if he had ever heard of Issa. He answered by saying that the Buddhists greatly respect Issa and that no one knew much

---

59. Prophet, *The Lost Years of Jesus*, 13n32.

60. Prophet, *The Lost Years of Jesus*, 13.

61. Prophet, *The Lost Years of Jesus*, 14n34 and 35.

about him except the chief lamas who had read the accounts of his life. The chief lama was reported to have said; "Also in our library we have manuscripts relating the life and actions of the Buddha Issa, who preaches the ancient tradition in India and also among the children of Israel. He was put to death by the pagans (the Romans). . . . When the holy child was still a boy, he was taken to India, where he studied to maturity the laws of the great Buddha."[62] When asked by Notovitch in what language the scrolls of *The Life of Issa* were written, the abbot replied that the documents pertaining to Issa were brought from India to Nepal and from Nepal to Tibet. They were written in Pali (a language derived from Sanskrit in which ancient Buddhist texts were written), but that a copy in Tibetan was in their monastery. Notovitch asked, "Are you committing a sin by showing these copies to a stranger?" The abbot replied, "I don't think I know exactly where the scrolls are in the monastery. But if you ever visit our monastery again, I will be delighted to show them to you." At that point, two monks entered the room, and the abbot was called away to the ceremonies.[63]

The next day, Notovitch left for Leh. He wrote that he would have liked to find a pretext to revisit the monastery. Two days later he had gifts brought to the head lama via a messenger, including an alarm clock, a watch, and a thermometer. He enclosed a letter expressing his wish to visit again. As he set out for Kashmir, his horse stumbled. Notovitch was thrown off and broke his right leg below the knee, which made it impossible to travel further. He was hoisted into the saddle with a provisionally splinted leg under tremendous pain; one servant led his horse while another supported his leg. After half a day of slow travel, he arrived again in Hemis. Everyone came out when the tour group arrived at the monastery.

---

62. Prophet, *The Lost Years of Jesus*, 15, 183–85.
63. Prophet, *The Lost Years of Jesus*, 185–86.

Notovitch wrote:

I am carried with great care into their best room and placed on a bed of soft materials next to a prayer wheel. The Abbot appears very pleased with his gifts. . . . Finally, at my request, he brings two large scrolls with yellowed leaves and reads Issa's biography to me in Tibetan, the translator translates what he says, and I record it in my travel diary. The curious document is written in the form of isolated verses, often without sequence. After a few days my condition has improved so much that I can resume the journey. Via Kashmir I travel back to India, slowly in twenty days, and the traveling causes me a great deal of pain. Via Srinagar I reach India before the first snow starts to fall. . . . For a long time, I pondered whether to publish the memories of Jesus Christ that I found in [Hemis].

The Documentary Jesus in India

In 2007 producers of the documentary *Jesus in India* visited the Jagannath Temple in Puri. The head of this temple, His Holiness the Shankaracharya of Puri, was one of the four most important Hindu leaders. He never gave interviews but, when he heard about the subject of the interview, he said yes. The crew was surprisingly allowed to interview him about Jesus's stay in India. Edward T. Martin was a member of the film crew and wrote a book about this experience and gives a summary of the interview. His Holiness began by noting that the biblical sources were silent about the period of Jesus between his twelfth and thirtieth years.[64] He said:

---

64. Edward T. Martin, *Jesus in India: King of Wisdom—The Making of the Film & New Findings on Jesus's Lost Years*, Reno, NV: Yellow Hat Publishing, 2008, 105–107; See *Jesus in India—The Movie* website.

The Jagannath Temple has a very ancient history as a center of learning and many important religious people have come to study in Jagannath over the centuries. However, these facts are swept under the carpet and although we in India are aware of Jesus' stay in India, the Christians are not willing to give any credence to our version. The fact that the Hindu tradition contributed to the education of Jesus is by no means accepted by everyone. Jesus studied here our teachings on the love of truth, grace, love, service, compassion, and ethics. It is also true that Jesus spent a long time in Kashmir, but that fact is also denied by the Christians. This can be read in a very old text, *The Bhavishya Maha Purana*, which describes a meeting between King Shalivahana of Kashmir and Issa Massi, Jesus the Messiah (after his crucifixion).[65]

The Shankaracharya added:

The Christians know that there is no information within their tradition about the missing years. Where was he during those eighteen years? Where did he live? Which countries did he visit? He lived in Kashmir and traveled all over India. But these facts are covered up and ignored.[66] (See An Asian Interlude.)

Was Yeshua in India after the Crucifixion?

The anonymous writer of *The Life of Issa* believed that Issa died on the cross. Yet there are numerous traditions among Hindus and Muslims that assume Jesus survived the crucifixion. A king of Kashmir, Shalivahana, was mentioned by the current Shankaracharya, the head of the Jagannath Temple, as

65. Hassnain, *Search for the Historical Jesus*, 191.
66. Martin, *Jesus in India*, 106.

having spoken with Issa when he arrived in Kashmir after the crucifixion. Shalivahana reigned in the first century.[67] Some people in India and Pakistan also believe that Yeshua traveled to the East after the crucifixion, having made his way to safety beyond the easternmost frontiers of the Roman Empire. He traveled around Central Asia again and would have died there at the age of eighty-one in the year 77 CE. I will come back to this information in chapter 13, about the events after the crucifixion.

67. Martin, *Jesus in India*, 106; Ahmad, *Jesus in Heaven on Earth*, 402, 409n1, 419, 424.

7

MARY MAGDALENE
IN HER YOUTH

The Hebrew name Miryam means womb. Her honorary name is "the Magdalene." This is derived from the Hebrew *magdal* or *migdal* for tower. This makes Mary's honorary name "the toweress" or "the exalted one." Both the proper name Miryam and the honorary title the Magdalene show a clear connection with the Great Mother of Ancient Israel.[1]

In the four canonical Gospels, Miryam played a varying role in the death and resurrection of Jesus—sometimes she appeared alone and at other times she was in a group of women. The three Synoptic Gospels mention Mary Magdalene as the first of the group of women and thus as the leader. John emphasized that she was the first to meet the risen Jesus. This shows that the early Christians highly valued and respected Mary Magdalene.[2]

The Gospel of John identified Mary of Bethany, sister of Martha and Lazarus, as Mary Magdalene. She lived in Bethany, close to Jerusalem.[3] John's view that Mary of Bethany and Mary Magdalene were one and the same is followed by many church leaders in the Western tradition.[4]

---

1. *MMU*, 73–78 is about the significance of the name Miryam and the title the Migdalah; see also *The Language of MA the Primal Mother*, 445; *The Black Madonna from Primal to Final Times*, 263.

2. Freedman, Myers, and Beck, *Eerdmans Dictionary of the Bible*, 865.

3. Jn. 11:1–5 and 17–45; Lk. 10:38–42.

4. Taylor, *The Coming of the Saints*, 31 names Tertullian, Ambrose, Jerome, Augustine, Gregorius, Beda Venerabilis, Rabanus Maurus, Odo, Bernard of Clairvaux and Thomas Aquinas.

However, the Eastern Orthodox Church assumed that Mary Magdalene and Mary of Bethany were two different persons.[5]

In the Western tradition another identification takes place, where Mary of Bethany, a.k.a. Mary Magdalene, is a sinner. The canonical Christian tradition has done Mary Magdalene a great injustice by putting her in the same category as the sinner of Luke's Gospel, who anointed Jesus's feet and dried them with her hair.[6] Luke's gospel states that Jesus was followed by "certain women who were healed of evil spirits and diseases: Mary called Magdalene, out of whom seven demons went out . . . "[7] An impression linking "sinner" to a woman from whom demons left led to the impression that Mary Magdalene must have sinned sexually, and the sinner became a prostitute. After that, Mary's reputation was tarnished for many centuries. The female sex became inferior and sinful; sex was connected with original sin. This tarnishing of women in general and Mary Magdalene in particular had been going on since the second century and culminated in the sermon of Pope Gregory in 591 in which he merged three women into Mary Magdalene: the sinner, Mary of Bethany, and Mary Magdalene.[8] Every effort was made to obscure the fact that she was Jesus's wife and the leading apostle of apostles in the original tradition. It was not until 1969 that the Church rid her of the stigma of being a prostitute. A further rehabilitation followed in 2016, when she was declared the apostle of apostles.

In many apocryphal texts she is the great initiate who guides others to higher knowledge and insight as the Migdalah or the toweress. The same image can be found in *The Gospel of the Beloved Companion*. In *The Gospel of the Beloved Companion* we hear the voice of Mary Magdalene; she called herself *Miryam*. I follow her example and often address Mary Magdalene as Miryam. *The Gospel of the Beloved Companion* made the connection between Mary of Bethany as the wife of Jesus and Miryam the Migdalah/Mary Magdalene. They are one and the same.

---

5. Prophet, *Mary Magdalene and the Divine Feminine*, 267.
6. Lk. 7:36–50; *MMU*, 310–11.
7. Lk. 8:103; Mk.16:9; *MMU*, 310.
8. Beavis and Kateusz, *Rediscovering the Marys: Maria, Mariamne, Miriam*, 20.

Miryam was with Yeshua at the Baptism in the Jordan.[9] They married each other in Cana.[10] She anointed him twice and was proclaimed the Migdalah at the second anointing.[11] From that moment on, Mary of Bethany was definitively Miryam the Migdalah/Mary Magdalene, and during the last meal she was appointed successor to Yeshua;[12] however, her leadership was not accepted by some of the disciples[13] (chapter 8).

The Alternative Sources

Anna of Carmel provided the following information through Claire Heartsong's channeling.

Miryam of Bethany grew up in the family home in Bethany and met Yeshua at a young age. Anna's account is consistent with Miryam's Gospel in that she said that Miryam of Bethany was one and the same as Mary Magdalene.[14] She was a daughter of Joseph of Arimathea and his second wife Mary of Magdala, whom Joseph of Arimathea married in the year 4 BCE fourteen years after the death of his first wife.[15]

Anna explained that Mary Magdalene's mother had a spiritual background. She was near Magdala, acting as high priestess in a cave where she performed the rituals for the Great Mother every year. Every year she took her eldest daughter, Mary Magdalene, with her. Mother and daughter were trained in the ancient matriarchal traditions, in which the *Shekinah*, or the feminine face of God, brought one to a direct inner revelation of God. Mary Magdalene learned the rituals, ceremonies, chants, rhythms, and cycles from her mother. The orthodox Pharisees and aristocratic Sadducees abhorred this tradition of the

9. *GBC* [4:1], 15; *MMU*, 96, paragraph 3: "Mary of Bethany is Mary Magdalene."
10. *GBC* [6.3, 7, and 9], 18–19; *MMU*, 113–25.
11. *GBC* [32.2], 59 about the second anointment; *MMU*, 353.
12. *GBC* [35:17 and 23], 65–67; *MMU*, 377 and 379
13. *GBC* [43:2–3], 81, *MMU*, 519–520.
14. Heartsong, *Anna, Grandmother of Jesus*, 148.
15. Heartsong, *Anna, Grandmother of Jesus*, chap. 25 "Mary Magdalene and Mariam's Childhood," 148–156.

Divine Mother. Because Joseph occupied a high-ranking position as an adviser to the Sanhedrin, the choice of his second wife compromised him. That was why his marriage to Mary of Magdala was shrouded in secrecy. Mary and her eldest daughter Mary Magdalene did not like to appear in public. They led a withdrawn life, with Joseph often absent because he was in Jerusalem or on a journey.[16] When Mary went with her mother to visit the surrounding villages, the hostile remarks of the Pharisees and Sadducees hurt her; they disliked women who worshipped the Divine Mother or the Shekinah. Rarely did mother and daughter leave their homes for nearby Jerusalem. Mary grew up in this climate of distrust; she opposed the dogmatic and hypocritical system that had such an impact on the relationship of her parents. At the age of nine, she met Yeshua and the spark immediately spread.[17] She visited the Carmel community several times and met Yeshua again.

At the age of thirteen, Mary was sent to Carmel for internal training. Yeshua prepared for a trip to England with his uncle Joseph of Arimathea, so he was not present during her stay. After all the luxury, comfort, and relative freedom in the parental home in Bethany, she had difficulty with the sober and disciplined existence in the Carmel community. She rebelled and was sent home. That same evening, she became seriously ill, after which her behavior changed radically. She was allowed to stay on probation until she was sixteen years old.[18] Then, she left with her cousin Mariam for Egypt to be trained as a priestess of Isis.[19]

The Regression Reports

Why were gifted Hebrew girls like Miryam and Mariam sent to Egypt by their families? Were there no training opportunities in Palestine?

---

16. Heartsong, *Anna, Grandmother of Jesus*, 148–150.
17. Heartsong, *Anna, Grandmother of Jesus*, 151.
18. Heartsong, *Anna, Grandmother of Jesus*, 152–56.
19. Wilson and Prentis, *The Essenes*, 179, 182; Wilson and Prentis, *Power of the Magdalene*, 76; Heartsong, *Anna, the Voice of the Magdalenes*, 366. Mary of Bethany goes to Egypt in the year 14.

There were basic training opportunities in Essene circles, but after basic training in the home communities, talented girls were sent to Egypt for advanced training. Remember that the mainstream climate in Palestine was discriminatory against women. According to the regression reports and certain historical sources, this is different with the Essenes, where women and men were equal to each other, but they were an exception in Palestine, where educational opportunities for girls were limited, so Miryam stayed in Egypt for a number of years to train as a high priestess.

In Egypt, the ancient tradition of the universal Mother was honored; there they called her Isis. In Upper Egypt, the Gnostic and early Christian texts from the Hebrew tradition of the First Temple have been found in the so-called Nag Hammadi library.[20] In it the Mother of Ancient Israel bears not the name of Isis, but Hokma or Wisdom (in Greek known as Sophia). Sometimes in these texts the Mother is also called by the following names: Pronoia; Protennoia; Eva; and Barbelo. All these names stand for (aspects of) the Great Mother. The climate in Egypt was much more liberal to women than in the strict and dogmatic Judaism of neighboring Judea, where the Pharisees predominated around the time of Yeshua and Miryam and interpreted the law to the petty letter. And it was in this climate that the Palestinian Essenes had to live and survive. They secretly remained faithful to the older tradition of the egalitarian and female–male image of God of the Father and the Mother.

The regression reports said that precisely because of her liberal Egyptian background and the powers she developed in energy transfer and healing, Miryam was distrusted and discriminated against after her return to Palestine. She was considered a stranger, and people feared her special abilities.[21] After returning from Egypt, she married her great childhood sweetheart: the Essene, Yeshua. She assisted him in every-

---

20. Source references to the Nag Hammadi library, or Nag Hammadi Codices, will be abbreviated NHC.
21. Wilson and Prentis, *The Essenes*, 179.

thing they undertook together in their public years. They had a common mission (chapter 8).

In mainstream Palestine this was not understood and strongly disapproved of, hence the negative press that Miryam received as a bride in the canonical scriptures. Her position was hard to digest even for the Essene priests of Judea. In their adaptation to Judaism and their emphasis on asceticism and celibacy, they had also become infected with the anti-woman virus. But their more liberal Egyptian background was not misunderstood. They gave Mary Magdalene the space to function fully—albeit in the background—as a healer and "co-teacher."[22] This image corresponds to Miryam's own Gospel, *The Gospel of the Beloved Companion*. The Essene followers understood her better because they understood and appreciated her Egyptian background. They saw that Miryam and Yeshua had a mission to fulfill together, in which they both needed to be supported. There is more information about Mary Magdalene's character and role during the public years in chapter 8.

AN ESOTERIC INTERLUDE: LIGHT CONCEPTION

From the regression reports from Anna of Carmel through Claire Heartsong, it was said that Mary Anna, the mother of Yeshua, was conceived from a light conception. The same goes for Yeshua, Mary Magdalene, and their daughter Sarah Anna. (There is more information about their other three children in chapter 15.) There were also other family members conceived in light conception. The regression reports also provide information about this subject.

Stuart Wilson and Joanna Prentis were first introduced to the concept of light conception when they read the book *Anna,*

---

22. Wilson and Prentis, *The Essenes*, 179. That Mary received her training as a priestess of Isis in Egypt is further elaborated upon in the two later books by Wilson and Prentis, which deal specifically with Mary Magdalene.

Grandmother of Jesus.[23] They were baffled, and they took the opportunity to question Alariel about this specifically. He gave a page-long lecture about light conception in relation to DNA. At the end he informed them that this topic was of interest to people of today because the children of the New Age, who come from other star systems, can only be born through a light conception.[24] The team questioning Alariel about this was joined in October 2006 by a biologist named Cathie Welchman; she was better able to ask Alariel specific questions about DNA because of her field of expertise.[25] While Welchman asked the majority of questions in that session, Prentis also occasionally asked a question or made a comment.

Alariel explained that in today's world, little is known about light conception. But the principle was known to esoteric groups in Yeshua's time and to mystery schools of the time. It was not as rare as it is now believed. Mary Anna, Yeshua, Mary Magdalene, and John the Baptist were all begotten by the Light. Among the disciples of Yeshua who were conceived by light: Sarah, the wife of Philip, and Mariam, the daughter of Rebekah.[26]

I came into contact with this subject earlier through the themes of androgyny, parthenogenesis, and virgin birth. In such cases, the female impregnates herself without the intervention of a male partner. Ancient priestesses are said to have possessed the ability of self-fertilization while in deep meditation. In the books of an old acquaintance and colleague from the international matriarchal studies network, Marguerite Rigoglioso, the themes of a miraculous conception and divine birth are further elaborated.

Alariel went on to explain that in a light conception, the mother provides the basic DNA; the father provides a blueprint

---

23. Wilson and Prentis, *Power of the Magdalene*, 133.
24. Wilson and Prentis, *Power of the Magdalene*, 134–143: 140, 142.
25. Wilson and Prentis, *Power of the Magdalene*, 135.
26. Wilson and Prentis, *Power of the Magdalene*, 135.

of consciousness on a subtle level and the Spirit does the rest. It is necessary to investigate the true nature of DNA, which exists on different levels, and some levels hold keys to consciousness rather than operating purely on a physical level. There are interdimensional layers of DNA that stimulate openness and awareness and subtle abilities, such as abstract thinking. These higher layers of DNA open the door to aspects of consciousness and make connections. Those higher levels of DNA that have not yet been detected by scientists are those layers of DNA that carry the consciousness of the father through the process of light conception. You see DNA as a transmitter of physical properties, but in reality it is a multidimensional information system. Your DNA contains not only physical attributes but also your total potential as an intellectual, cultural, and spiritual being. It forms the blueprint for your development. It's like a growth map, showing your potential on all levels. The lower the level at which the DNA functions, the greater the input will be from the parents. Physical DNA contains a lot of input from the parents, while multidimensional DNA, which functions at high levels, reflects the qualities and expressions of the soul.[27]

Sarah Anna was spiritually the child of Yeshua and Mary Magdalene; genetically she was the child of Mary Magdalene alone, because only one set of physical DNA, the set that comes from the mother, was involved in this kind of light conception. So those who believe that descendants of Yeshua live on Earth with his physical DNA have not understood this. The Spirit integrates the DNA so the child does not develop health problems, as is the case with modern cloning. By the working of the Spirit, the health and constitution of a child conceived by the Light is much better and stronger than most other children.[28]

27. Wilson and Prentis, *Power of the Magdalene*, 134–35.
28. Wilson and Prentis, *Power of the Magdalene*, 134, 137.

It is well known that two double strands of DNA are active in the process. The question now is how do you produce two double strands when you genetically only have one double strand from the mother. That's what the Spirit does. This creates the second double strand that curls around the mother's. In doing so, the Spirit descends through many frequencies to the physical level and moves ever deeper into matter. The Light crystallizes and condenses itself until it manifests in the substance of the physical DNA. The Spirit does not create the DNA, The Spirit becomes the DNA. The light descends into the physical level. The male provides higher-level DNA, the consciousness-level DNA, not the physical-level DNA. On the physical level, Spirit produces the second double strand of DNA. There are small changes to strengthen the DNA and not allow diseases from the mother's side. The DNA is optimized. Thus, a child is conceived who is a stronger and better person.[29]

Welchman asked, "How can the non-present physical male DNA still give physical characteristics to the baby; otherwise, it would be a clone and completely like the mother?"

Alariel replied, "The Spirit becomes the child, but the father participates in this because he provides an imprint of consciousness through his spiritual partnership and cooperation with the mother; by living with her and not by a simple act. He delivers high and subtle levels of DNA throughout the process of coexistence and spiritual partnership."[30]

Alariel continued: "When an advanced soul desires to experience human life on Earth, that soul vibrates at too high a level to embody through the usual mating process with a physical body. Beings who resonate with desire are born from a desire-based process. But beings beyond desire cannot be born this way, because the process does not conform to their

29. Wilson and Prentis, *Power of the Magdalene*, 136–37.
30. Wilson and Prentis, *Power of the Magdalene*, 137–38.

consciousness. To come into incarnation, a being needs that vibrational frequency that reflects the vibration of its own consciousness. Therefore, advanced high-vibration beings can only embody in an advanced, high-vibration way."

Welchman then asked, "Why are not all human beings born of a light conception?"

Alariel replied, "They haven't earned that right; not yet. They need to become high-vibrating beings, and most people have not yet achieved that. But many New Children who come from other galaxies do. They can only be born here through the process of light conception. Soon there will be many more light-begotten children, and so it is good that we look into this subject."

Mary Magdalene
in the
Public Years and After

Figure 8.0. Nicholas Roerich, *Madonna Protectoris*, 1933.
The Nicholas Roerich Museum in New York. Mary Magdalene is
considered by many researchers as the Ma-Donna or
Notre Dame of France (see chapter 18).

8

MARY MAGDALENE'S MISSION

What picture of the adult Mary Magdalene do the alternative sources provide? As discussed earlier, after Stuart Wilson and Joanna Prentis wrote their first book, *The Essenes*, they collected more material from regressions and wrote two books about Mary Magdalene with help from an unexpected source: the angel Alariel.

Mary Magdalene's Role in Modern Times

After Prentis thanked Alariel for the chance to speak with him, she started to ask Alariel why his group of angels were interested in the Essenes.

Alariel said, "We are interested in all Melchizedek operations, and the Essenes were particularly significant because they lived at an important time for your planet. The cycle downwards into material density had to be brought to a close, and the upward spiral into the Light had to begin. The Essenes played a major part in this whole process. They were the main focus of Melchizedek activity on your planet at that time. The consciousness and the technology of the Essenes were quite advanced for their time. They had an early form of electricity, something that your culture regards as a very recent invention, yet two thousand years ago the Essenes lit their dwellings with electric lamps."[1]

---

1. Wilson and Prentis, *Power of the Magdalene*, 16, refers to Laurence Gardner, *Lost Secrets of the Sacred Ark*, and gives a diagram and a description of the Baghdad battery that was able to produce light.

Prentis asked about the Essenes now in incarnation. "What is their main task?"

Alariel answered, "Partly they are here to support a reassessment of how things were two thousand years ago, particularly in the group around Yeshua. They are here to encourage all people of goodwill within the Judeo-Christian tradition to take a look at the whole narrative of Yeshua's life in a much broader way. In doing so, they may be able to build a bridgehead between the people actively working in the Christian churches and the more progressive and liberated people, for whom new ideas and new perspectives seem quite natural."[2]

To the objection that "new ideas and new perspectives" are a major challenge for many Christians, Alariel replied: "Change has always been present in the Christian experience, and there are aspects of Christianity that bring the promise of renewal and fresh beginning. In particular, there are a number of groups that are now focusing on the testimony of the female disciples and the central importance of Mary Magdalene. Mary's story has enormous potential to move people out of their old, rigid ideas and into a new understanding of the role women played at the time. This not only provides a refreshing new perspective, but it also starts to build an 'alternative history' to put it alongside the traditional accounts. In time this may even lead to a reassessment of Yeshua and his teachings."[3]

Prentis asked, "Which element of human consciousness that we have lost is important for us to bring back now?"

Alariel answered, "Undoubtedly, it is the Goddess element. The Sacred Feminine is the source of the highest and most subtle wisdom and the most profound knowing. Shamanic cultures have always recognized the primacy of feminine wisdom, and it is time that this primacy—symbolized by Mary Magdalene—is more widely acknowledged in the word. It is only by acknowledging the primacy of feminine wisdom that conflicts can be resolved and that the deep wounds that

2. Wilson and Prentis, *Power of the Magdalene*, 18.
3. Wilson and Prentis, *Power of the Magdalene*, 18.

humanity has inflicted upon itself over the centuries will finally begin to heal."[4]

Mary Magdalene and Yeshua Together Form a Transforming Vortex

Alariel went on to report the following about Yeshua and Mary's energies combining in a powerful way.

When they met again on his return from India, Yeshua had become a fully fledged initiate in the Ageless Wisdom, so he had reached the level High Priest. Mary had risen to become a High Priestess in the Isis tradition. These two advanced and enlightened beings came together at that time as spiritual equals in a partnership of profound significance for the world, and being on these levels their energies had become remarkably focused and powerful. When you reach that level, your consciousness automatically generates a twelve-pointed transformational vortex around you, a vortex of change so powerful that it will cause ripples of change in the consciousness of all those who are anywhere near you. . .

When two vortices of transformation are in the same proximity, the interconnection of energies boosts the combined power, so they form an energetic focus so powerful that—to the visions of the angels—they appear as a great beacon of Light visible throughout the entire galaxy. Now you can begin to see why Mary and Yeshua, working closely together at the time of their Ministry, had such a powerful effect on those they encountered. When Yeshua spoke to a large crowd of people and Mary Magdalene was present, the divine masculine provided structure and focus while the divine feminine provided the energy to boost and sustain the process. When Yeshua spoke to a crowd, even though Mary said nothing, she was a vital part of the process, since her transformational vortex, linked

---

4. Wilson and Prentis, *Power of the Magdalene*, 24.

with Yeshua's, provided a complete communication system. . .

That communication carried in words, in energy, and in mudras (special hand gestures, the principle of which Yeshua had learned in India and developed in his unique fashion), which provided a multilevel expression. Even if you were deaf, the energy and the mudras give you the whole message. And even if you were deaf and blind the energy of these two combined transformational vortices could reach you and change your life forever. . .

When Yeshua was speaking to an audience, his hand gestures supported the spoken message, but he was also working energetically on higher levels. Many of those who watched him speak noticed that his eyes seemed to focus just above the heads of his audience. As he looked out over the assembled crowd, Yeshua could see the architecture of consciousness rise up in a semitransparent form above the head of each individual. Within each form that the consciousness projected, there were similar colors and almost identical patterns of movement. Within the architecture of each consciousness-form he could see similar areas of blocking and resistance to change and similar patterns of restriction and rigidity. Yeshua was able to work simultaneously on several levels to release the blocks in the consciousness of all those who were present. Because the crowds were often tightly packed around him, leading to an overlapping at the auric level, this meant that the wave of transformation could spread through the crowd with the rapidity of a forest fire. And as blocks were released and a more natural openness and flow were restored to consciousness, Yeshua could see the result in shifts in color frequency and changes in form and pattern.[5]

When someone in regression remembered to have been James, a younger contemporary of Yeshua and Mary, he told how the group of refugees was later visited in Gaul by Mary Magdalene. "She was so beautiful, she had dark eyes. She was our link with the *Kaloo* [the ancient

---

5. Wilson and Prentis, *The Magdalene Version*, 57–60.

ones from Atlantis and cofounders of the Essene order], and she brought messages from them. It was she who was informing our scattered people of the jobs we must do. She was telling us that it is through the female line that the emotional aspects of the memories will be held."[6]

It was the first time that James met Mary. Prentis understood from his body language that it made a deep impression on him. When asked about Mary's energy field, James replied: "It was huge, purple and blue." Now James became so emotional that he could hardly utter a word. "She linked with many other beings, on the Earth and off the Earth; it was all tied in with Light, so you feel as if she was bringing a vortex of the whole universe's memory into your place." In response to a question from Prentis, he said emphatically that she did not create a vortex: "She was a vortex."[7]

The Essence of the Teaching:
Love and Forgiveness to Reach
Connectedness and Unity

"What was Yeshua's purpose?" To this question from Prentis, Alariel replied: "What he was trying to do was to establish a new spiritual path, a new way of living and relating to people and to God, using the Cosmic Energy of Love, which he was anchoring into the Earth reality. He was encouraging people to focus that energy through the heart center. This energy is profoundly transformative. It moves through the dry forest around the heart like a cleansing fire, sweeping away the dead wood of past experiences and all the anger, fear, and hurt that people cling to. When all that has gone, it empowers the individual in expanding awareness and rising into higher frequencies of consciousness. All this reestablishes the Children of Light in a new relationship with God."[8]

When asked about the central theme of Yeshua's teaching, Daniel

---

6. Wilson and Prentis, *Power of the Magdalene*, 38.
7. Wilson and Prentis, *Power of the Magdalene*, 38–39.
8. Wilson and Prentis, *Power of the Magdalene*, 19.

Benezra replied: "The need to reconnect with the heart energy, the energy of unconditional love. Yeshua not only talked about this energy, but also demonstrated it in his life and his being. He gathered and focused the love energy, and because his energy was so strongly within him, others were able to feel it and express it in their heart. This great cosmic energy of unconditional love, focusing through Yeshua and through his being, made it available to everyone. To love the person in front to you—whether they smile at you or snarl at you—was quite a new idea for them. Yeshua understood that we are all children of the Creator and is it not natural to love the children of a loving Creator? He understood the nature of God, because that nature was essentially love and was expressed as love. He accepted everyone as a follower, even the greatest sinners, including anyone who left their past behind and turned towards the Light. Yeshua lived in the moment, and when someone came to him truly repenting of the past, Yeshua lifted him or her into the moment, and it was as if the past ceased to exist for them."[9]

Sometimes Daniel joined the group of followers around Yeshua who moved from village to village. Daniel said:

I could not travel with them as much as I would have liked, having responsibilities as an elder in my community in Hebron. I made it a point of going with them as often as I could. From the very beginning when the group began to gather around Yeshua, I knew that this was something very special. And when I was there I felt so much more alive than I usually did, so it made me determined to go with them whenever I could. It was so light, and everyone was so happy to be around him. . .

I remember he healed a man who had been blind for many years. Yeshua laid his hands upon him and when he took his hands away, he told the man to open his eyes. And he did so, and he could see. And this man had a child and he had never seen the face of his son, and now he was able to look upon the face of his own child. It was

9. Wilson and Prentis, *The Essenes*, 148–150.

a wonderful moment and there was a wonderful feeling in that village. It was as if healing was taking place for all those who were there. Not only were the bodies healed, but also the doubt and fear and disbelief were healed as well. It was as if the whole pattern of the universe had been leading up to that point, and the lives of all the people present in that village had been leading to that. It was a big opportunity for everyone who was present. Many people had a rebirth in their hearts, so there was healing on many different levels.

This was a central theme of Yeshua's teaching to come back into original Oneness. Forgiveness is a vital part of embracing Oneness, and that is one reason why Yeshua taught this. When you completely and utterly forgive someone, what they have done to you or those dear to you, this trauma ceases to exist in your consciousness. It is not that you stop holding this event against them, but that for you the event doesn't exist anymore. As it no longer exists, it no longer separates you from the doer of this deed, hence it is not an obstacle to unity. This level of forgiveness is hard to achieve, but Yeshua told us it is a key to entering Oneness and it should be the aim of all those who follow the Way.[10]

The Marriage Between Mary Magdalene and Yeshua

Daniel Benezra explained: "I sensed that there was some deeper mystery in their relationship. When they were together within one of the inner groups there were moments when great energies seemed to focus through them, giving the simplest movement a special power and grace. I find it difficult to describe this effect, for this was so subtle. I can only say it was like watching the mysteries unfold like a flower opening or the performance of some ancient and sacred dance. It was very subtle and that is why I find it difficult to explain. This only happened in the inner groups, when only the disciples and some Essenes were present."

10. Wilson and Prentis, *The Essenes*, 146, 151, 282.

The more Daniel saw Jesus and Mary together, the more he marveled at this mystery.[11]

Prentis asked Alariel: "There has been much speculation recently about whether Yeshua and Mary Magdalene had been married. Were they?"

He replied: "Yeshua and Mary were spiritual partners. Yes, they did go through a marriage ceremony—the Wedding at Cana—but they were not man and wife in the conventional sense. They had important work to do together, and marriage was simply the most effective way of achieving this within the rigid social customs of the time."[12]

Alariel frequently emphasized that they were spiritual partners. When asked where the opposition to this marriage came from, Alariel gave the following answer: "The tendency to deny that Yeshua was married, let alone married to a controversial figure like Mary Magdalene, initially came from deep within the Essene order. At that era it was accepted that a rabbi was married, but Yeshua was not a mere rabbi; he was the Essene Teacher of Righteousness. The Essene priests, some of them almost obsessed with the concept of purity, insisted that their wonderful and long-awaited Teacher of Righteousness should be pristine and separate, above and beyond all earthly things. The lay brothers, on the other hand, were much more open to the truth of a real relationship between Mary Magdalene and Yeshua. Knowing that even amongst their own ranks they were so deeply divided about this, the Essenes decided to keep the marriage secret, even to the non-Essene disciples. The Essenes were very good at presenting a unified front to the world, and they did this most effectively. This attitude spread to other parts of Jewish society, so it became taken for granted that Yeshua never married. It is only now, after all this time has passed, that the truth is coming out and that the real role of Mary Magdalene—and the debt which the world owes her—is becoming publicly acknowledged."[13]

Prentis asked: "Was it essential for spiritual reasons that Yeshua and

---

11. Wilson and Prentis, *The Essenes*, 185.
12. Wilson and Prentis, *Power of the Magdalene*, 96.
13. Wilson and Prentis, *Power of the Magdalene*, 97.

Mary Magdalene worked closely together?" Alariel responded:

Yes, completely essential. The energy required to initiate a big breakthrough needed to be balanced before it appears in full manifestation
upon the Earth. Anchoring the cosmic energy of love—unconditional
love—upon Earth was a major project and needed powerful and balanced energy. No single human could have accomplished that task.
The crucifixion process was part of this, but just one part of it.
Another part was the life Jesus and Mary shared, blending and align-
ing their energies and forming a single Star of Oneness, an energy
vortex that created a portal through which this Love Energy could
descend fully and anchor itself upon the Earth. This energy had long
been available to the advanced few, but humanity as a whole could
not access it before Yeshua and Mary anchored it through their balanced focus. If this energy had not been securely anchored on Earth,
those who came after Yeshua and Mary would not have been able to
lock onto it in their consciousness and so could not have applied this
energy in their lives. The vibration of cosmic love would then have
been too subtle for their consciousness to sustain and hold. . .

The great outpouring of this spiritual energy of Love affected
the whole evolutionary process of human development. If you look
at the arc of human history, there was a steady increase in the density of physical existence starting in Lemuria, accelerating during
the Atlantean period and reaching a high point of physical density
(and a corresponding low point of spiritual sensitivity) during the
Roman Empire. The point at which Yeshua and Mary were born
was the very nadir of spirituality on this planet, the lowest point
of the devolutionary arc. Because Yeshua and Mary Magdalene
moved human consciousness forward into an upward spiral, this
opened up many new opportunities for human beings to rise in
vibration and access higher levels of consciousness, subtler frequencies of light and being, that would not have been possible within
the previous downward spiral. By turning the devolutionary
spiral into an upward evolutionary one, Yeshua and Mary saved

humanity from a long period of existence at a much lower level. Your planet would have plunged ever more deeply into darkness and despair. The collaboration of Yeshua and Mary saved humanity from all that and in this aspect of their work the whole world has reason to be grateful to them. They saved humanity from the possibility of future sins, not the burden of past karma. Yeshua and Mary Magdalene were the combined saviors of humanity and should be honored as such.[14]

There is a recurrent theme of Mary the mother as the co-redeemer or in Latin the *co-redemptrix*; some people in the Catholic Church wished to formulate this in a new Mary dogma. But according to Alariel, that honor belonged to Mary Magdalene. Or, in my view, possibly to the collective group of Marys that included Mary the mother and Mary Magdalene.

Daniel reported "although Yeshua was far beyond any of us, yet I saw that Mary Magdalene was more his equal than we were. Of course, both the masculine and feminine energies are aspects of the greater divine energies, and as Essenes we knew the symbolism of the Father and the Mother very well. We were used to a concept of the feminine presence. Shekinah is the Presence of God in the world."[15]

Elsewhere, Alariel asserted that:

In the early days of Christianity it was not understood that the whole Universe is balanced from Father-Mother God downwards. Because this was not understood, the role of women as leaders and wisdom bearers could not be understood, and powerful women seemed to be a threat to the whole patriarchal structure of the Church. So how do you deal with an empowered woman? You marginalize and malign her. You say she is a prostitute and hope that no one will pay attention to her.

---

14. Wilson and Prentis, *Power of the Magdalene*, 98–99.
15. Wilson and Prentis, *The Essenes*, 185, 196.

Mary Magdalene worked with Yeshua to anchor the energy of unconditional love into the vibrational fabric, the energy field of the Earth. Their life together and the balancing of their energies as they worked side by side made this possible. This has been forgotten—or to be more exact, has never been understood. Her part in this was not at all recognized. She did this by being Yeshua's spiritual partner and working closely with him. And by putting up with all the negativity that came from the patriarchal Jews around her, while still remaining in that center of unconditional love. . .

Mary and Yeshua demonstrated a new kind of partnership between two advanced beings. Their work on anchoring the energy of love into the matrix of the Earth laid the foundations for all future expansion of the light on this planet. Through this new model, you can begin to glimpse a better way forward, with man and woman working together in a greater harmony where each respects the different talents and abilities of the other. This leads to a rethinking of the whole basis of relationships. Their partnership reflected the ultimate balance of Father-Mother God, which lies at the heart of the Universe.[16]

This is the core of the ancient gnosis, as I was once taught in early gnostic studies.

Prentis asked Alariel: "Did Yeshua and Mary Magdalene have any children?"

Alariel replied, "Ah, this is a key question. We will return to this at a later stage."[17]

The Female Disciples Had Two Leaders: Mary Anna and Mary Magdalene

According to Alariel, Mary Anna, the mother of Yeshua, and Mary Magdalene had different characters. About Mary Anna he said:

---

16. Wilson and Prentis, *Power of the Magdalene*, 100, 146–47.
17. Wilson and Prentis, *Power of the Magdalene*, 97.

Mary Anna is a person of quiet and calm authority. There was a great sense of clarity about her. She acted as the spiritual anchor and foundation for the group of first circle female disciples, being well qualified to do so as she had been a high initiate since the days of Akhenaten, in Egypt. She was one of the brightest jewels in the main Egyptian mystery school of that time and achieved a balance and spiritual empowerment that made her the ideal leader for the first circle of female disciples. . . Mary Anna could have taken ascension back at the time of Akhenaten when he presided over the main mystery school in Egypt about 1300 [years] before Yeshua was born. Mary Anna was a very advanced being but has not been given credit for this by the Church. She could have taken ascension long before, but she stayed to do this specific work. . .

Some people mistook the quietness of Mary's manner for docility or weakness, but when they looked into her eyes, they saw a quality of steely and unmovable determination that surprised them. A few people might have been foolish enough to start arguing with her but that look of determination made them change their minds very quickly. . .

When she led the meetings of female disciples, Mary Anna started the discussion, but from that point on she said very little. She let everyone have their say and expressed themselves fully, and only when a matter has been thoroughly discussed, she rounded this off with a balanced and moderate summary, which acted as a fair consensus for the whole group. . .

Mary Anna was definitely the spiritual anchor and foundation of the group of the first circle of female disciples. She had the most experience of this whole group, but Mary Magdalene was nearly upon her level, so there were two very advanced initiates heading up the group of female disciples. . .

Mary Magdalene had a different temperament; powerful, eager, and enthusiastic. The fire of her total commitment to the light burnt within her like a bright flame. She was a passionate advocate of truth and justice, but in her zeal, she may have occasionally overstated the case, something Mary Anna would never do. If Mary Anna's empowerment expressed itself in her poise and restraint, Mary Magdalene's empowerment manifested in a confident and joyful outpouring of her energy and love.[18]

Daniel shared the following:

Through her link with Yeshua, Mary Magdalene was able to ascend through many levels of consciousness into a state of being which some of the disciples saw was far beyond their own position and which they envied greatly. This envy was the cause of a good deal of resentment. . . Mary understood the Kingdom of Heaven better than any other of the disciples; far better in fact, for the only one who remotely approached her in understanding was John, apart from Mary the mother of Jesus. Mary Magdalene was . . . followed at a distance by John. He knew how advanced Mary Magdalene was and he was glad to have her counsel. The more conventional disciples, however, were too proud to ask Mary for advice. . .

Mary and John were very close and those two were the two most beloved of the disciples. Mary taught the disciples especially when Yeshua was away, for he often took time on his own to commune with the Spirit in quiet places. Mary was in truth the supreme disciple, who became the teacher of the teachers, because she understood the path to the light so well. She came into her own after the crucifixion, when many of the followers turned to her. Some were too proud or stubborn to listen to Mary, but I think that was their loss.[19]

---

18. Wilson and Prentis, *Power of the Magdalene*, 90–1, 101–02; see also Wilson and Prentis, *The Essenes*, 274.

19. Wilson and Prentis, *The Essenes*, 184, 186–87; see also *Power of the Magdalene*, 101. Alariel states that Mary Magdalene had many followers under the female disciples. The Pistis Sophia also states that Mary's heart is focused on the Kingdom more than the other disciples.

The Female Disciples
Next to the Male Ones

Alariel reported the following about how the groups of disciples were organized:

It is important to understand that the discipleship system that Yeshua set up was designed to mirror the greater symbolism of the Universe. . . The balance of the Father-Mother God is mirrored in a balance of male and female circles. There are six circles of twelve, making 72 male disciples, and six circles of twelve, making 72 female disciples; a total of 144 disciples in all. The names of the first circle of twelve male disciples have come down and are well attested.

. . . In the case of the second circle of male disciples you have, for example, Joseph of Arimathea, a very major player, but having so many other duties, it was not possible for him to be a first circle disciple. . .

The first female circle is composed of the following persons. First of all we had two advanced initiates, who were frankly head and shoulders above the rest, and these were Mary Anna, the mother of Yeshua, and Mary Magdalene.

- You have Helena Salome, who took the matronymic name of Mary. Outside the family group she was often called Mary Salome. She was the sister of Mary Anna and therefore in the biblical account of the crucifixion she is simply called Salome. She married Zebedee and her sons John and James were disciples.
- A fourth member was Mary Jacoby, sister of Mary Anna and aunt of Yeshua. She married Clopas, also called Cleopas (or Alphaeus). I add to Alariel's information that their sons, James the Younger and Levi Mattithyahu or Matthew were disciples too.
- Their daughter Abigail was Yeshua's niece; she married John, the disciple. Abigail was also part of the first women's circle, together

with several other nieces of Yeshua, such as Mariam Joanna, Sara, Lois Salome, and Susannah Mary.[20] In addition, there were Laura Clare, younger sister of Yeshua, and Martha of Bethany and her sister.[21] Of this first circle of twelve, Mary Anna, the mother of Jesus, was the leader. Mary Magdalene was, if you like, the second in command in this group.

---

20. Mariam Joanna is a cousin of Yeshua, daughter of Mary Anna's sister Rebekah; Sarah is daughter of Isaac, who is brother of Mary Anna, and she marries the disciple Philip; Laura Clare, or Ruth, is sister of Yeshua and daughter of Mary Anna; Lois Salome is daughter of Joseph of Arimathea and niece of Yeshua; Susannah Mary is daughter of Joseph of Arimathea and niece of Yeshua.

21. The sister referred to here is called "Mary" and by this is meant Mary of Bethany. In Wilson and Prentis's first book, *The Essenes,* and in Heartsong's first book, *Anna, Grandmother of Jesus* (chap. 25), Wilson and Prentis and Heartsong still assume that Mary Magdalene and Mary of Bethany are one and the same and that she is the wife of Yeshua. However, Heartsong's second book *Anna, The Voice of the Magdalenes,* which was published twelve years after her first book, contains new information. There you'll read that Yeshua's first wife was Miryam of Tyana and that Yeshua would later become engaged to Mary of Bethany, his second wife. Heartsong claims on page 106 of *Anna, The Voice of the Magdalenes,* that Mary Magdalene is a composite figure of three persons: Miryam of Tyana, Mary of Bethany, and Mariam of Carmel, niece of Yeshua. These three Marys are—esoterically—the mainstays of the Magdalene order and are spiritually fully integrated with each other. In *The Magdalene Version* (p. 55–56) Wilson and Prentis follow these new insights from Heartsong and distinguish between Miryam of Tyana and Mary of Bethany. I state here emphatically that for me Mary of Bethany is Miryam the Migdalah and Mary Magdalene. Confusion arises because there were several Magdalenes who were initiated into the esoteric order of the Magdalenes. There is a difference between an initiation title and a personal name. In *The Magdalene Version,* (p. 55), Stuart and Prentis admit that this causes confusion. There are several Magdalenes but there is only one Mary Magdalene. The Magdalenes are initiated in Egypt into the Isis Mysteries and master the mystery knowledge about dying as ending the separation and rising in one's own inner light. They act in the way Isis raised Osiris and impregnated herself with Horus. In my opinion, the female (and supportive male) Magdalenes could be the Hebrew branch of the Egyptian Isis mystery schools within the Essene order. When the Magdalenes incarnate, their job is to carry the Isis frequency further out into the world. To be a Magdalene thus indicates a degree of initiation. The fact that many in the twenty-first century are attracted to the Magdalenes is because they recognize this Isis vibration and then propagate it further in a snowball effect. This is important in the context of the increase in the frequency of light on Earth.

Several decades before the birth of Yeshua, a conference was held in the Interlife, the time-space between lives, presided over by the Archangel Michael. This conference was attended by human beings not in incarnation, and by a number of angels, especially angels who guided and assisted those souls preparing to incarnate. Human beings who were in incarnation at that time attended the conference during a sleep state. The object of this conference, firstly, was to gather and focus those volunteers who wished to assist in the forth-coming life and work of Yeshua. These volunteers included many souls who had worked with Yeshua during a number of lifetimes and could be said to constitute the "soul family" focusing around him. All spiritual teachers gathered a soul group of supporters who often wished to incarnate whenever their teacher happened to be on Earth.

The second object of the conference was to plan how these souls might come together around the teacher in a certain time and a certain place: that was Israel two thousand years ago. The angels started to investigate suitable families for the incoming souls. So these souls would have reached adulthood at the time that Yeshua began his teaching work. The overarching aim was the building of a strong team of disciples around Yeshua, especially those in the first and second circles of disciples, both male and female. The connections between the female disciples were given the greatest attention of all for it was recognized that these disciples would have the hardest task. As you have seen from the list of first circle female disciples, they were all related to Yeshua, either directly or through marriage. All these connections served to fuse the first circle of female disciples into a single dedicated and tightly knit team whose fierce loyalty and intuitive linking enabled them to think, feel, and act as one. . .

As isolated women in a rigid society, they would simply not have been taken seriously. So, their close family connections, resulting from close soul connections developed through many lifetimes, were a necessary part of ensuring that they could do their work. Thus, a potential weakness in the group was turned into a strength through the hard work of many angels setting up this grouping and the good-will and outstanding heart quality of all within the group. If you are looking for a good demonstration of the Way that Yeshua taught, you might cite the harmonious collaboration within this group of women . . .

The male disciples, containing some relatives, but coming from different areas of Israel never achieved this degree of closeness and coordination, nor did they need to. Their task was to present a diverse spectrum so the Way could never be dismissed as only a maverick branch of an extended Essene family. The power and com-mitment of the male disciples within the first circle was such that they would have been taken seriously, even if they stood alone.[22]

So much for the record of this angelic assembly. This background information shows how carefully the first women's circle was put together. Let's continue with the second circle of female disciples.

Alariel said:

The second circle of twelve female disciples were enormously sup-portive of the first circle, and they provided support in covering family and community duties, which made it possible for the first circle to spend much time with Yeshua and Mary Magdalene. Yeshua regarded the help and generous, open-hearted sharing amongst the female disciples as a model of Unconditional Love.

Yeshua brought this to the attention of the male disciples, amongst whom it was not widely understood or very well received. They regarded themselves as teachers and leaders and saw the role of women as listening and following. So, it was hard for them to accept that the females had anything to teach them. Yeshua put Unconditional Love at the very center of his teachings; this made it all the more irritating to the men, who would have preferred some

---

22. Wilson and Prentis, *Power of the Magdalene*, 78–9, 82, 84–5, 87–8, 101.

more intellectual and philosophical doctrine based upon reason rather than upon heart energy. They felt ill at ease in matters of heart energy and longed to elaborate or structuralize Yeshua's teaching in some way, to make it more intellectually challenging and therefore more impressive to the male mind. They saw heart energy as much too vague and insubstantial and longed for a set of logical and profound concepts so they could seize the high ground of intellectual debate. They saw Yeshua's great intellectual gifts in debating (especially with the Pharisees be encountered), but they could not understand why he chose so flimsy a foundation for his teaching as Unconditional Love.[23]

At a young age I learned that the knowledge of the heart is at the center of the gnostic tradition, but in those early days, I was intellectually oriented and did not really understand the meaning of this. Therefore, these words from Alariel touched me and made me—after decades of study—understand what the knowledge of the heart really meant.

Alariel continued:

Of course they dared not raise these questions while Yeshua was teaching them. . . The boldest among the non-Essene male disciples only went so far as to mutter amongst themselves that Yeshua was started to feminise the tough, heroic, patriarchal core of Judaism by taking such an extraordinary line in his teaching . . . as promoting a feminine weakness in the Judaic consciousness. . . . Bear in mind that the male disciples were brought up in a Jewish culture that still believed in an eye for an eye and a tooth for a tooth. The leap from that position to turning the other cheek and loving your enemy was simply too big a step for them. . .

The male disciples were drawn from several areas of Israel, and some were Essenes and some not. This was deliberate so no one

---

23. Wilson and Prentis, *Power of the Magdalene*, 80–81.

could dismiss the teaching as a Galilean movement or an Essene cult.
. . . The male disciples were independently minded individuals and
you could not say they were a close-knit team. There was no single
universally accepted leader among them. There was a conventional
faction, led by Peter, and a progressive faction including Thomas,
James, and Philip. There was also John somewhere in the middle
trying to keep the peace, trying to pull everything together so they
can function as a group.[24]

As Alariel described it, the situation in the women's group was completely different:

The female disciples were bound to have a more difficult task than
the male disciples because they were coming into a society that was
so rigid and patriarchal. To survive that and still be able to work
and teach—how difficult that was going to be. The women formed
a close-knit group, a group which had been working towards this
moment during the centuries. Over many lives they had been gaining
experience at various spiritual levels. They had been meeting at various mystery schools, they had been coming together and planning
what they would do. So this group had been working on this project
for a very long period of time, which is why they choose to assem-
ble in a very close-knit extended family group, many of them being
born about the same time. Their families were interknit and closely
connected, so there was an instinctive trust and solidarity amongst
the female disciples that simply wasn't there in the case of the male
disciples. The women were all telepathically linked and formed one
complete telepathic unit. In addition to this link, this closeness and
mutual trust, they had the advantage of having one leader, Mary
Anna. When Mary spoke, they all listened. She was the matriarch
figure, a very substantial person of quiet authority. So, if a meeting
needed pulling together, Mary did this. If a difficult decision had to

24. Wilson and Prentis, *Power of the Magdalene*, 81–3.

be made, Mary focused the group and saw that this was done.[25]

The closeness of the women's team did not please the male disciples. They did not react well to a group of twelve women all moving with one intent. They found that quite eerie, quite frightening. The male disciples were wary and suspicious from the beginning. Twelve women who all moved as a unit with only one goal, the men found that creepy and frightening. They did not understand it, and that was another motivation for them to marginalize the women disciples, and another reason to write all this out of history. The story of the female disciples is a big story, and its time has come. Various members of this close-knit team will be stepping forward quite soon to tell their part in this story. Yes, their time has come. Other groups are now working on this as we speak. And much has yet to be told in this story. It's not only the first circle of women disciples, but the second and to some extent the third circle as well. Many things will flow from the information from the female disciples as they come forward.

. . . It is time for a shift in human consciousness. And the empowerment and leadership role of women is an essential element in leaving the past behind and beginning to access the full potential and creativity available to the human race as a whole. It strengthens the recognition that God is balanced masculine and feminine. The whole system of male and female discipleship, as set up by Yeshua, reflected the central balance of Father-Mother God. Any system that honored the Father energy and neglected the energy of the Divine Mother lacks flexibility, sensitivity, compassion and wisdom. It will tend to become rigid and brittle and will fragment into a number of competing elements. It is one of the great tragedies of Western culture that all the knowledge of the female disciples of Yeshua had been lost and now is the time to restore that knowledge.[26]

---

25. As leader of the first women's circle, Mary Anna keeps a connection with the Melchizedek system, in which the frequency of unity, peace, and light are passed on. Mary Magdalene is described as a strong shining star.

26. Wilson and Prentis, *Power of the Magdalene*, 83–85; see also Wilson and Prentis, *The Essenes*, 274.

Contested Leadership

After the crucifixion, Mary Magdalene's leadership was contested. According to Alariel, "Yeshua had intended his disciples to set up a twofold structure, the basic structure. After Yeshua had left them, the teachings would be given out in two ways. The outer teachings would be spread by most of the male disciples, led by Peter, who was to be the rock, the foundation, of this new movement. Whilst the inner teachings—the Inner Mysteries of the Way—would be taught by John, James, Thomas, and Philip, with this group led by Mary Magdalene. Mary saw the two teaching arms as mutually supportive because the outer group would be open and public and would deflect attention from the inner, which in any case needed a quieter environment to do the more subtle work along esoteric and gnostic lines. Mary believed that in time the most advanced followers of the Way would move from the outer to the inner group."[27]

Alariel reiterated that Mary Magdalene was the leader of the mystery school of Higher Knowledge. Her power was by no means recognized and acknowledged by all the disciples, although they had all witnessed the fact that Jesus and Mary Magdalene had deep conversations, which some of the disciples already did not feel comfortable with. Alariel said:

They felt fear at the power of Mary. And the fact that she was a priestess of an Egyptian cult, something that would be regarded by the Pharisees as an alien and heathen cult, did her no good whatever in the eyes of conventional Jews. The Pharisees looked down on her with fear and loathing. Fear because by going through the Isis initiations, she became a living Goddess. Are not patriarchal men always frightened of a Goddess? So, they only had the weapons of darkness to use against her with their lies and distortions. Both Mary Anna

---

27. Wilson and Prentis, *Power of the Magdalene*, 92; Wilson and Prentis, *The Magdalene Version*, 70.

and Mary Magdalene were advanced initiates who were leading the group of female disciples. Jesus had a partner of great spiritual attainment. How can any Christian organization deny leadership roles to women, when one woman was the spiritual partner, closest collaborator, and chosen companion of its founder?[28]

Perhaps it is superfluous to say that the picture painted here with Mary Magdalene in a leading position, which is not acknowledged, agrees with various gnostic sources and with *The Gospel of the Beloved Companion.*

Peter's Aversion to Powerful Women

Alariel said:

Peter, who had trouble dealing with any empowered woman and who had real difficulties with Helena Salome, also called Mary Salome, Mary Anna's sister, had problems with Mary Magdalene from the very beginning. Her free and powerful expressing of opinion, which might have seemed frank, open and engaging to a person, grated upon Peter's patriarchal nerves. Whilst he would never dare to question Mary Anna—one look from those eyes would have silenced him—questioning, and indeed criticizing Mary Magdalene was something Peter often did.

. . . Peter deeply resented her closeness to Yeshua and wished to promote the interests of the male disciples and diminish the importance of the women. He could not bear the thought that Yeshua told Mary Magdalene things that he—the obvious leader (in his eyes at least) of the male disciples—needed to know. Above all, Peter could not accept the basic structure through which the Way would be spread.[29]

---

28. Wilson and Prentis, *Power of the Magdalene*, 100–102.
29. Wilson and Prentis, *Power of the Magdalene*, 91–92.

During a meeting immediately after the crucifixion and resurrection, James and Mary Magdalene tried to explain the twofold plan to Peter. He brushed it off at once as impractical. He saw the followers of Yeshua as an embattled army; splitting meant weakening their forces. Moreover, Peter could not bear the thought of a rival leader—a rival leader who happened to be a woman was quite unthinkable for him. He saw this move by James and Mary Magdalene as an attempt to weaken his leadership, and he simply would not tolerate this.[30] In taking this position Peter began the rift between the mainstream teaching of the Way (which in time became the Catholic and Orthodox churches) and its gnostic counterpart. This led to the final persecution and elimination of all gnostics as heretics, a persecution which paradoxically was carried out by the very part of the movement—the outer Church—which had been designed to nourish and protect it.

The outer aspect of the Way developed into the Church, but the inner aspect was a spiritual movement of free kindred souls. It was a movement in which spirituality and not religion was the driving force. As Islam had the Sufi movement and Judaism had the mystical Kabbalah, Christianity should have also had its own mystical and gnostic core, and the loss of this core left it with a permanent wound from which spiritually it never recovered. Christianity was damaged so deeply that it could not fulfill its purpose as Yeshua had planned. Without its balancing mystery school counterpart, the Church is like a clock with some of its mechanism missing. You have not really experienced Christianity yet, only the incomplete fragment that history has handed down to you. That fragment has helped and inspired countless people since its foundation, but it could have been infinitely more effective as a way of treading the spiritual path that leads to the Light.[31] Further:

The balance between the outer Church and the inner mystery school . . . was also a reflection of the essential nature of the universe. The

---

30. Wilson and Prentis, *Power of the Magdalene*, 92; *The Essenes*, 179–81; *The Magdalene Version*, 70.
31. Wilson and Prentis, *Power of the Magdalene*, 93–94.

outer Church represents Father God, and also the Sun, knowledge, structure, ritual, and form. The inner mystery school presents
Mother Goddess and also the Moon, wisdom, flow, process and life.
Neither is complete without the other and together they manifest
the wholeness of the Way. The balanced path of spiritual development has been rejected in favor of a one-sided presentation that valued the form, but rejected Life.

. . . In fact, the people of that time, except for a few, were unable
to convert the fear into love in their lives; humanity was not yet
ready for it. Therefore, Christianity developed along conventional
religious frameworks. Today, it is very different because large groups
of people are awakening.[32]

Yeshua knew that Peter would have the greatest difficulty accept-
ing the key role of Mary Magdalene and would have probably tried to
isolate her and bar her from the inner group around him. But Yeshua
wished the group to move toward the time of crucifixion as united as
possible. He hoped that the shock of these traumatic events would make
Peter accept a larger role for Mary Magdalene.[33]

On many occasions Yeshua talked with John and Mary Magdalene
alone. They knew his innermost feelings and the full extent of his philosophy. John and Mary shared their Essene upbringing, they were connected by family ties, and both were members of the core group.[34] They
grew in spiritual stature under Yeshua's guidance. They developed a
strong mutual bond of friendship and respect. They were both Yeshua's
beloved disciples and the most advanced of the disciples. They were
further than most conventional followers, such as Peter and Andrew.[35]
During those conversations, Yeshua urged John and Mary to appreci-
ate Peter, because he spoke the language of the common people and
knew how to express himself in words that the average person could

32. Wilson and Prentis, *Power of the Magdalene*, 94–95, 103–04.
33. Wilson and Prentis, *Power of the Magdalene*, 93.
34. Wilson and Prentis, *The Magdalene Version*, 68–69.
35. Wilson and Prentis, *Power of the Magdalene*, 66; Wilson and Prentis, *The Essenes*, 186.

understand.[36] Yeshua also asked John to be a peacemaker between the men's and women's groups; between the outer group led by Peter and the inner group led by Mary Magdalene. He asked him to be an intermediary and bridge, and John accepted.[37]

Andreas and Peter were jealous of the close relationship that Mary and John had with Yeshua. They also distrusted both John and Mary, because both had a much more subtle understanding of spiritual matters than (the fishermen) Andrew and Peter. In addition, both felt resentment because they fell outside the group of Yeshua's favorite disciples. John enjoyed great respect from the other male disciples, and they did not dare to attack him. Mary Magdalene was an easier target despite the fact that she had a large following among the female disciples.[38]

When asked why Yeshua withdrew after the crucifixion, Alariel replied that it was necessary for him to leave Israel. He could not stay because many people wanted to harm him, and he could no longer teach openly. Via Cyprus, he secretly left for Damascus and traveled eastwards on the old trade route. The route took him through cities that today are called Baghdad, Tehran, and Kabul. It was a long journey before he finally reached India. Before leaving for Damascus, he spent time in Cyprus, where Joseph of Arimathea owned a large estate. He took a rest there and recovered from the crucifixion before undertaking the long journey to the East. Joseph had made Cyprus the center of trade in that region, a wise decision given the rapidly changing situation in Israel. From Cyprus he controlled the shipping of tin to all Roman Mediterranean ports.[39]

When asked why Yeshua did not accompany his immediate family and followers, including Mary Magdalene, to England and France, Alariel said that he knew once the teachings were given, the teacher must step

---

36. Wilson and Prentis, *The Magdalene Version*, 67–68.
37. Wilson and Prentis, *The Magdalene Version*, 65; Wilson and Prentis, *Power of the Magdalene*, 83.
38. Wilson and Prentis, *Power of the Magdalene*, 67, 101.
39. Wilson and Prentis, *Power of the Magdalene*, 106.

back; this way the disciples can develop the work further. Then the teacher should disappear. This put full responsibility on the next generation, which is free to continue to grow and explore things in their own way.[40]

Prentis asked, "What convinced the disciples to preach openly when they knew it was dangerous? Was it the resurrection or was it the teachings of Yeshua?"

Alariel replied, "It was partly Yeshua's teaching, but the 'Something else' was the descent of the Spirit. It was a breakthrough that many of the disciples made their connection with the Spirit at that time which motivated them to preach openly. In the energy of Spirit, anything is possible. This energy was intense at that time, because so many disciples were stepping up and becoming teachers in their own right. Although it was a chaotic and traumatic time, it was also an exhilarating time because the Spirit was so much present."[41]

Asked if the message of Yeshua had come true through his resurrection, Alariel responded: "Yeshua spoke Truth, and when someone speaks Truth, you know it in your heart. You feel the energy of Spirit, and it does not require a miraculous event to show you that what has been spoken is true. However, in the cultural climate of those days, it was perfectly understandable to link the message of Jesus to the resurrection. Many religions at that time contained elements that today would be described as magical. It was widely accepted that the teaching would be confirmed by miraculous events surrounding the teacher. Therefore, it was quite natural that the early Christians used the resurrection as a magical proof for Yeshua's teaching."[42]

From Resurrection to Ascension

Prentis remarked that the resurrection had become an important part of the Church's teaching. She cited as an example of Yeshua's

---

40. Wilson and Prentis, *Power of the Magdalene*, 107.
41. Wilson and Prentis, *Power of the Magdalene*, 107.
42. Wilson and Prentis, *Power of the Magdalene*, 108.

statement: "I am the resurrection and the life,"[43] to which Alariel replied, "It is important to understand that the Bible was divinely inspired, but humanly edited. The original document read, 'I am the ascension and the life,' but an editing scribe thought the word *ascension* was a mistake and substituted resurrection for it. The sentence that follows culminating in words 'will never die' makes it clear that ascension was meant here. This is because every physical body, even a resurrected body, must eventually die, while a Light Body, the body of ascension, is immortal. Resurrection only gives you access to continued existence in a physical body, whereas ascension gives you access to Eternal Life. Would you rather have a longer physical lifespan, or as the Essenes put it, 'Eternal joy in a life without end'?"[44]

Prentis asked in which form Yeshua appeared to his disciples after his crucifixion, to which Alariel answered, "After the crucifixion, Yeshua reached a level of development where bilocation—something he had practiced before—became significantly easier for him. That was why he kept in touch with his mother, with Mary Magdalene and all the key disciples on a regular basis through bilocation and talking to them."[45]

Prentis asked, "Do the people of two thousand years ago know about bilocation?"

Alariel replied, "Only the initiates of the mystery schools fully understood the process."

Prentis then asked, "How should we deal with this when almost every church has a stained-glass window of the crucifixion over the altar?"

Alariel responded, "Let me help to picture you a church of the future. When you enter it is completely light, and above the altar there is a beautiful stained-glass window that shows Yeshua and Mary Magdalene, hand in hand. Shown symbolically above them in the sky there is the higher resonance of this partnership which is

43. Wilson and Prentis, *Power of the Magdalene*, 108. The phrase is from John 11:25; it is not found in Miryam's gospel.
44. Wilson and Prentis, *Power of the Magdalene*, 108–109.
45. Wilson and Prentis, *Power of the Magdalene*, 109.

Father-Mother God. No pain, no suffering, only the essential balance of the Universe."[46]

The Significance of Mary Magdalene

To Prentis's question about the real meaning of Mary Magdalene and her relevance today, Alariel said:

> The real importance of the Magdalene is her role as an empowered and enlightened woman. She stands forth as a continuing inspiration to women all over the world. Your world is now moving towards a state where each of you is becoming your own sovereign and your own priest or priestess. The key question is where the authority resides. The teachings of Yeshua put authority firmly within the transformative power of the Spirit. In one of the summaries given by the angels of Yeshua's teachings, it said: "The Spirit of God, working its miracle of change within the heart, becomes the ultimate authority, the agent of transformation and the arbiter of Truth." Within the group around Yeshua, Mary Magdalene was the clearest carrier of that Truth. And that is why Yeshua called her "the woman who understands the All." She had been one of the most successful of his disciples, a true teacher. . .
>
> The early Church Fathers were happy to accept Mary as the first "witness of the resurrection," but they wanted to confine her to a witness role and prevent her from becoming recognized as teacher. She was allowed to only be a witness, an apostle to the apostles, but not to the world. The Church had already started to develop Christianity from its beginning as a God-centered presentation, steadily changing it until it became a Savior-centered presentation focusing on a Divine Savior-Hero. The concept of a Divine Hero is familiar in Hellenistic and Roman culture, but quite alien to the Judaic tradition. The Jews perceived their long-awaited Messiah as a king in the David line, a great prophet and the restorer of the glory

---

46. Wilson and Prentis, *Power of the Magdalene*, 109.

of Israel, but not as the Son of God. The Church started to blur the boundaries that separated the human Yeshua from the Divine Jesus. In that process of deifying a Divine Jesus, Mary Magdalene became an obstacle, because as his spiritual partner she drew attention to his humanity. By emphasizing the divinity of Yeshua, the Church could play the role of intermediary and gain power and prestige. Therefore, Mary must be defiled and falsely accused of being a prostitute. It is one of the greatest injustices of the Western world and it's only now, after all those years, that this injustice is being exposed.[47]

Prentis said, "It certainly seems that Mary Magdalene's ability to influence and inspire is increasing rather than diminishing."

Alariel replied, "The power of the Magdalene does not diminish or wither with the passing of time. She is the Wayshower, embodying the energy and wisdom of the Sacred Feminine and inspiring others to do the same. She held a unique position amongst the disciples of Yeshua and she demonstrated in her life the transforming power of the Spirit. Mary Magdalene was able to attune to the power of the Spirit and that gave her the most complete understanding of the All. The male disciples might be able to understand the All intellectually, but she absorbed the All into every level of her being and therefore she became the ideal living pattern of what discipleship should be."[48]

The Rise of the Gnostic Movement

According to the alternative sources, the Essene order flourished from about 160 BCE to 30 CE. It is also stated that from 30 CE and especially during the Jewish revolt from 66 CE onward the Essenes merged with the rising gnostic movement. This information from an esoteric angle corresponds to the historical material and to modern insights.[49]

---

47. Wilson and Prentis, *Power of the Magdalene*, 144–46.
48. Wilson and Prentis, *Power of the Magdalene*, 146.
49. Freedman, Myers, and Beck, *Eerdmans Dictionary of the Bible*, "The Essenes," 426: "The Essenes departed from history sometime after the war with Rome."

What exactly do the alternative sources report? According to Daniel Benezra the core group, supplemented by the elders from various subcommunities, met immediately after the crucifixion. Mary Anna introduced a senior member of the Melchizedek order on Earth and announced that he would preside over the meeting. This person informed them that after the crucifixion the outward and visible role of the Essene order in the world had come to an end. He thereby announced its dissolution.[50] When asked about the rise of the gnostic movement, Alariel said:

The Gnostics were scattered throughout the Middle East, but there was a big concentration of Gnostic groups in Egypt, the land where many Gnostic texts have been discovered. Within Egypt, the groups in Alexandria were particularly advanced and influential. They could trace their roots back to the large Essene community in this area. Although the Essene order was dissolved in the last quarter of the first century, an oral tradition still lived there. Generations of Essenes in Israel and Egypt sent their brightest daughters to the temple of Isis in Alexandria.

. . . This was a very broad-based movement with major differences between the Christian and Jewish Gnostics and a variety of views even within these two branches. Some elements of Gnostic thinking predated the rise of Christianity, making it difficult to structure this movement into rigid categories. Gnostics were often classified according to their theological ideas by the hostile Church Fathers and also in modern handbooks, but that is not correct. The Gnostics were concerned with the process of enlightenment and ascension, a form of self-liberation. They call that process the Way. To walk the Way was a process.

. . . For Gnostics, theology is only a series of mind-games that their most brilliant thinkers liked to indulge in. It was a diversion and not the heart of things. For the mainstream Christian followers of Peter and Paul, the theology of doctrine was the core of their

50. Wilson and Prentis, *The Essenes*, 300; Wilson and Prentis, *The Magdalene Version*, 50.

faith. The Gnostics reasoned differently. If process could take you into enlightenment and ascension, what value would you put on theology in comparison to that? What is the point of crossing theological swords when only the process can take you to enlightenment and ascension? To self-liberation?

There is no general agreement on theology amongst Gnostic thinkers. Theologically, there are many different groups of Gnostics: the liberal and the conservative. Yes, some of them supported the dualistic creation featuring the Demiurge, but progressive Gnostics like Mary Magdalene certainly did not, and there is no trace of dualism in her statements, quite the reverse. In fact there is a continuing returning to the theme of Oneness.[51]

These comments touched my heart. This thinking in Oneness is, in my view, the oldest form of Christian gnosis. It appears—as I explained in an earlier book—in the Nag Hammadi library of Upper Egypt in various first-century gnostic writings. It is important to note that the oldest gnostic writings reflect Oneness and not a dual creation by a lesser, malevolent divine being, the Demiurge.[52]

The Christian Mystery School Led by Mary Magdalene

When Yeshua was no longer physically present, many people fell back on Mary Magdalene. According to Alariel, "She was the Keeper of the Flame of Truth, and it seemed perfectly natural to them to turn to her for guidance and counsel. Mary was widely respected as a source

51. Wilson and Prentis, *The Magdalene Version*, 64, 63, 65–66.
52. van den Broek, *Gnosis in de Oudheid*, 48: "There is now almost general agreement among researchers that the Gospel of Thomas does not belong to the radical, mythological gnosis, because nowhere does it presuppose the fall of Sophia and the wicked creator god."; 91: "But while the older Gnostic works are keenly interested in the origin of our world of ignorance and death, and how man may be delivered from it, little attention is paid to it in the text Zostrianus (and others)."

of secret Wisdom, and for the progressive gnostics she was becoming a beacon of inspiration in what was becoming an increasingly difficult and dangerous world."[53]

Now that Yeshua was not physically present, it was up to Mary Magdalene to set up the Christian mystery school and to give shape to the divine feminine in the person of Spirit or Wisdom. According to Alariel, she organized the school on three levels.

1. Local: one meeting was held per week
2. National: people met four times a year at a central location to exchange ideas and learn from each other
3. International: they met once a year at the estate of Joseph of Arimathea in Cyprus.[54]

Alariel explained to Prentis:

Joseph of Arimathea provided the transport for Gnostics attending these gatherings through his network of ships and provided a communication transport centered in the office at his estate. Through messages carried on the ships, this office kept all the emergent Gnostic groups in contact with each other. Although the intention was to gather once a year during midsummer, there had been eight meetings in eleven years. This had to do with the busy schedule of Mary Magdalene, who also occasionally traveled to England to teach the priestesses of Avalon.[55] Joseph of Arimathea himself was rarely present because of his busy work elsewhere.[56]

The participants, travelling on Joseph's ships, arrived from various locations. Mary Magdalene arrived from the Languedoc. John came from Israel and later from Ephesus. Philip traveled from Greece. Only the most progressive, open-minded Gnostics attended

---

53. Wilson and Prentis, *The Magdalene Version*, 73.
54. Wilson and Prentis, *The Magdalene Version*, 71.
55. Wilson and Prentis, *The Magdalene Version*, 71–72, 79.
56. Wilson and Prentis, *The Magdalene Version*, 73.

the meetings. They came from a range of backgrounds, including Hellenistic, as well as Jewish-Essene, but they shared a passionate commitment to the Truth. The more conventional Gnostic groups, especially those close to the emerging mainstream Christian Church, prefered not to attend the gatherings. They knew that Mary Magdalene had taken a leading role and they were uncomfortable with that format.[57]

Joseph and his attendants made sure that the guests lacked nothing in Joseph's cool and well-furnished villa. The estate was an oasis of calm in the midst of their lives full of uncertainty and tension. They could relax and develop long-term plans. The participants developed a theoretical and practical basis for the most advanced and progressive part of the Gnostic movement. The Midsummer gatherings were essentially the result of the collaboration between John and Mary Magdalene. John chaired the meetings and Mary gave the central speech on a subject that had been agreed in advance as the central focus for that year's gathering. Afterwards, deliberations continued for several weeks.[58]

Yeshua had always felt the weight of Jewish tradition upon his shoulders, with the Pharisees waiting for him to say something that they could use against him. Mary Magdalene was much freer during the Midsummer gatherings. She could speak freely and spoke from a direct source of inner wisdom. Her compassionate heart won her followers amongst all the groups who heard her speak. At last, on the island of Cyprus she could speak all that was in her heart and mind, for here she was among colleagues and friends.[59]

During each of her eight speeches, Mary explored one major theme, examining the subject thoroughly, but concisely. That is why these speeches were later called The Summaries. They were never circulated to other groups and that is why no written form of them

57. Wilson and Prentis, *The Magdalene Version*, 74.
58. Wilson and Prentis, *The Magdalene Version*, 75.
59. Wilson and Prentis, *The Magdalene Version*, 73–74.

has survived into the present era. But our angelic group has access to the Akashic Records, and this group has worked for some time to recover them and convert them into modern language. In Cyprus, Mary enthusiastically seized the opportunity to explain her philosophy. When the last Gnostic died in France in the third century, much of this secret knowledge died with him. Her philosophy continued to be taught in Mary's extended family in the Languedoc. But the teachers within this tradition became so scarce that they were unable to protect the purity of the transmission.[60]

This led to mistaken ideas, like the Dual Creation, creeping into the tradition by the time of the Cathars, who retained the practice of the Way, but their knowledge of the philosophy of Mary was largely lost. The pure transmission of the Melchizedek Wisdom via the Essenes and Mary Magdalene's teaching came to an end. At the end of the planetary cycle, the veils of forgetting are dissolving and knowledge that had been lost for many centuries is being restored. With humanity moving though transition, it is right and just that access to it should be restored to you, as it may prove helpful at this time.[61]

I will only transmit here the titles and core of Mary's eight teachings.

1. She spoke in the first teaching about three trinities. At the root of everything is the power of the Law of Three.[62]
2. In the second teaching, Mary spoke of the sacred Universe, created by God the Father-Mother according to the principle, as above so below.[63] There is no inferior part in this.
3. The third teaching was about the soul. The soul travels through the Universe; it leads from the Light, via the shadow world, back to the Light.[64]

---

60. Wilson and Prentis, *The Magdalene Version*, 75–76.
61. Wilson and Prentis, *The Magdalene Version*, 77.
62. Wilson and Prentis, *The Magdalene Version*, 83–86.
63. Wilson and Prentis, *The Magdalene Version*, 87–90.
64. Wilson and Prentis, *The Magdalene Version*, 90–93.

4. The fourth teaching dealt with salvation as an achievement of the enlightened consciousness itself; salvation was not affected by faith in the savior or teacher, which shows only the Way to salvation. We must save ourselves.[65]

5. In the fifth teaching, she outlined the Way as the process by which the Christ Energy transforms one's life and consciousness.[66]

6. In the sixth teaching, she explained how the Universe moves continuously toward harmony and how it is maintained by a system of balanced forces and counterforces that ultimately converge at a center of balance, wholeness, or oneness.[67]

7. In her seventh teaching, Mary spoke of truth and peace. Either you want to find the truth of things, or you don't. If you are committed to the light, to the truth, you will let go of everything until you find it.[68]

8. The eighth and last teaching was about Oneness and the All. When you live in harmony with the All, you are a friend of all and an enemy of none, because you perceive Father-Mother God in everything you see.[69] This is at the core of Mary Magdalene's philosophy.

By the time the last meeting was held, it was clear that the lives of these gnostics were getting progressively more difficult. Most of them realized that was the last opportunity for the whole group to meet. Mary gave a farewell speech, which Alariel quoted in her words:

We are the ones who know the joy of love, the joy of union. We are the ones who know the true connection with one another. We are not separate. We recognize this connection as love. There are no boundaries for we know that love is infinite. We are aware of one

---

65. Wilson and Prentis, *The Magdalene Version*, 94–96.
66. Wilson and Prentis, *The Magdalene Version*, 97–103.
67. Wilson and Prentis, *The Magdalene Version*, 104–108.
68. Wilson and Prentis, *The Magdalene Version*, 109–112.
69. Wilson and Prentis, *The Magdalene Version*, 113–119.

another as pure light, pure love, and pure consciousness. It is so easy to be love and to hold this energy.

Now we know that we must move into other realms that are not so full of love, yet if all is well, these too may be raised up into the light. There is so much beauty in the physical realm. Mother Earth is so beautiful. There is so much life here and we are so deeply connected to Mother Earth, as she is connected to Father Sun, our shining star. At a deep level of experience, all is One, and the sign of this Oneness is the great gift of the love that connects us all.

It is so good to know that I have a spiritual family who walks with me, a family who knows the truth. With this family I can be who I really am. It is so good to remember this time together, and when we go out into the world, we will carry the energy of love with us. When we all meet again, perhaps in other lives wearing other faces, we will recognize this golden energy and know that we have connected again with the great family of love.

So, go out now into the world with courage and with joy, taking the energy of love with you. Let this energy spread out like golden ripples until all the whole world is blessed and unified.

May love fill your hearts always, and may the angels bless you and guide you upon your path.[70]

70. Wilson and Prentis, *The Magdalene Version*, 137–39.

PORTAL THREE

THE GOSPEL
OF THE
BELOVED COMPANION

BY MIRYAM THE MIGDALAH

Figure 9.0. Mary Magdalene teaching the disciples.
From the St. Albans Psalter (England) from 1119–1123, plate 13b, 51.

9

THE GOSPEL OF THE BELOVED COMPANION

In 2010 a new Mary gospel emerged; it was translated and published by Jehanne de Quillan. The full title reads *The Gospel of the Beloved Companion: The Integral Gospel of Mary Magdalene*. In the last sentence of her final chapter, the author reveals her identity and states: "I am Miryam, called the Migdalah, the Beloved Companion."[1] In 2021, I published a commentary about this gospel in my book *Mary Magdalene Unveiled*. I deeply investigated the authenticity of this Mary gospel.

Jehanne de Quillan provided the following background information about Miryam's text: "It is generally believed that the text was brought from Alexandria to Languedoc, then Roman Gaul, during the early to middle part of the first century. Originally written in Greek, it was first translated into Occitan during the early part of the twelfth century. This particular version has remained preserved within our families and spiritual community since that period."[2]

For her English translation, de Quillan used the original Alexandrian-Greek source text. She aimed to keep the modern English and French translations as close as possible to the original text. In the original document, like all ancient texts, the sentences were written

---

1. *GBC* [44:1], 82.
2. *GBC*, 4; *MMU* 5.

consecutively, without a break; the punctuation as well as verse and chapter divisions were added later by the translator. In her book, she compared the Alexandrian Greek of Miryam's text with the original Greek texts of the canonical Gospels and commented on this. In addition, in her comparison she also included fragments from noncanonical, apocryphal gospels such as those of Thomas, Philip, and the Gospel according to Mary. The similarities are so striking that you get the impression that various apocryphal texts have also borrowed elements from the Greek text of Miryam; in other words, Miryam's text seems to be a source text.

Jehanne de Quillan shared that the spirituality of the Miryam's gospel had permeated the lives and hearts of members of her community for many generations. The intention of the 2010 publication was the hope that Miryam's gospel will once again speak to the hearts of people all across the world.[3]

The First Dedication to the Cathars

It is not difficult to guess why the first dedication in Jehanne de Quillan's book in *The Gospel of the Beloved Companion* is addressed to the Cathars. In one of the next chapters I will prove that when Mary Magdalene and her family left for the south of France after the crucifixion, she took her text with her. In France the text was handed down from antiquity to the Middle Ages. In the twelfth century the text fell into the hands of the Cathars. They consider it "their" gospel. They've hidden it and handed it over—at the risk of their own lives—for nearly one thousand years into the twenty-first century. Many medieval Cathars were tortured and lost their lives; the first devotion to the Cathars, found on the third page of *The Gospel of the Beloved Companion,* is obvious:

---

3. Personal mail contact with Jehanne de Quillan in September 2020.

To the Burned, the Slaughtered,
The Tortured and the Persecuted.
For all who have suffered and all who have died
At the hands of Darkness and its Inquisition,
But never has the Darkness overcome them.
After 700 years the Vine turns green again.[4]

Jehanne de Quillan immediately informed the reader of the great sacrifices her Cathar community had to make to hide, preserve, and pass on Mary's text for about ten centuries, from generation to generation, in the deepest secrecy. The last sentence, "After 700 years the vine turns green again," indicated a rebirth of the Cathar heritage. It is a reference to the vision of the Cathar Guilhem Bélibaste.

In 2010, Jehanne de Quillan and her Cathar community decided, after long and difficult mutual deliberations, to return the secret *Gospel of the Beloved Companion* to the world. The ranks were divided to the core in the decision whether to release it because they still feared being the target of aggression from "the dark ones." That is why the document's whereabouts were kept secret as well as, to this day, no photographs of the document have been issued.

The Cathar Church is the church of the female Holy Spirit, and the symbol of the Spirit and the Cathar Church is the dove. A peace monument was erected in the French town of Minerve in memory of the 140 Cathars who were burned at the stake there at the foot of the fortified town in 1210. Exactly eight hundred years later, in 2010, Miryam's gospel was published. The monument is called *La colombe de lumière* or the *Dove of Light*. Carved into a stone column, it was created in 1982 by the sculptor Jean-Luc Séverac. The question has always been: How did the medieval Cathars gain their knowledge of Wisdom or Spirit? How did they know that Jesus and Mary Magdalene were married?

---

4. *GBC*, iii.

Figure 9.1. *La colombe de lumière* or the *Dove of Light*, carved into a stone column, by the sculptor Jean-Luc Séverac in 1982. Photo by Jaap Craamer.

The Cathars, Mary Magdalene, and John

That the French Cathars knew Jesus and Mary Magdalene were married is evident in a statement about Cathar Christians from Albi, mentioned by one of their opponents, who remains anonymous. The text said: "They also learn at their secret meetings that Mary Magdalene was the wife of Christ."[5] In 1959, the Dominican scholar Antoine Dondaine

---

5. The 13th century Cistercian monk and historian Peter de Vaux de Cernay wrote that Mary Magdalene was Christ's "concubine": "Further, in their secret meetings they say that the Christ was born in earthly Bethlehem and crucified in Jerusalem, was wicked and that Mary Magdalene was his concubine and that she was the woman who had committed adultery, in which the scriptures refer." A document possibly written by Ermengaud of Béziers, undated, anonymous, and appended to his *Treatise against the Heretics*, makes a similar remark.

published a medieval treatise that said the Cathars regarded Mary Magdalene as the wife of Jesus.[6]

Many apocryphal texts have circulated among the Cathars. According to the French historian of religion, Henri-Charles Puech, a Cathar such as Bélibaste knew large parts of the Gospel of Thomas by heart.[7] This Gospel of Thomas was only excavated at Nag Hammadi in 1945, but Bélibaste already knew about it in the Middle Ages. He would have quoted from it in conversations with his inquisitors.[8]

The crusader army started the Albigensian War with the siege of the southern French city of Béziers. They opened fire on July 22, 1209, the name day of Mary Magdalene. It was an unparalleled massacre. The crusade lasted twenty-five years and had a devastating effect on the south of France. According to Jehanne de Quillan, more than one million people died.[9] The land of the Cathars was destroyed and their numbers decimated. Guilhem Bélibaste predicted that after seven hundred years the laurel or the vine would bloom again, and the era of the Spirit will begin—symbolized by the dove, the symbol of the Cathar Church. One of the fruits on that blossoming laurel or vine must be *The Gospel of the Beloved Companion* written by Miryam and handed down in a secret Cathar community for more than ten centuries.[10] In this gospel Miryam describes a female Holy Spirit.

Jehanne de Quillan wrote that the Inquisition records of the thirteenth and fourteenth centuries of hundreds of Cathar initiates, also

---

6. Antoine Dondaine, "Durand de Huesca et la polemique anti-cathare," *Archivum Fratrum Praedicatorum*, tomus 29, Rome: Istituto storico domenico, 1959, 228–276; *The Black Madonna*, 320n79; *MMU*, 51n2.

7. Henri-Charles Puech, *En quête de la Gnose*, Paris: Gallimard, 1978, partie I: la gnose et le temps, partie II: l'Évangile selon Thomas, partie II, 51n1, 56. About logion 114 see Johann Joseph Ignaz von Döllinger, *Beiträge zur Sektengeschichte des Mittelalters*, 2 vols, New York: Burt Franklin, 1890, vol II, 244, 245, 164–226. About logion 30 see Puech *En quête de la Gnose*, 251 and von Döllinger *Beiträge*, 210; *The Black Madonna*, 320n76; *MMU*, 51n4.

8. Bram Moerland, *Schatgraven in Nag Hammadi*, The Hague, Netherlands: Synthese, 2004, 32; *The Black Madonna*, 320n77.

9. *GBC*, 7.

10. *GBC*, 8.

called parfaits and parfaites, revealed their unusual devotion to the Gospel of John.[11] According to the inquisitors' reports, the "heretics" accept only the Gospel of John as a true testimony of Jesus. In other words, they don't accept the other three canonical gospels. It is highly probable that the Cathars regard *The Gospel the Beloved Companion* as their own version of the Gospel of John—as John's original source document.[12] As mentioned earlier, however, the author of this source document was Mary Magdalene. Copies were in circulation in Palestine and elsewhere. She took her text to France. But her transcripts continued to circulate elsewhere. In later time they were "adapted" to masculine standards, reflecting the patriarchal resistance and the growing backlash against female leadership.

Jehanne de Quillan argued that, since the assumption of Church Father Irenaeus in the second century, it was common practice to identify the male "beloved disciple" in the Fourth Gospel as John.[13] Many people today assume that references to the Fourth Gospel of the Beloved Disciple are indications of John. However, as the information was given by Cathars interrogated during the Inquisition hearings, these were in fact references to *The Gospel of the Beloved Companion*. Nevertheless, the inquisitors mistakenly believed them to be references to the Fourth Gospel of John.

It now has become clear that numerous contemporary theologians believe that the beloved disciple of the fourth gospel may have been Mary Magdalene.

The Uniqueness of the Gospel of Miryam

Jehanne de Quillan spent three years working on her book, but then the authentic Alexandrian-Greek text was removed—in complete secrecy—to a place unknown to her. Even today she does not know where the text is hidden. There is no photograph of the original manuscript and the name

---

11. *GBC*, 174.
12. *GBC*, 175.
13. *GBC*, 175.

of the place where it is kept is unknown. The name of the group that keeps it is unknown. In my book *Mary Magdalene Unveiled* I researched the reliability of Miryam's document. I have examined it in detail and compared it with the text of the Fourth Gospel of the Beloved Disciple or the Gospel of John. I found that there are many similarities in chronology and structure between the two texts. Entire paragraphs are literally the same. Yet there are essential differences. What are they?

1. Central in Miryam's Gospel is Wisdom or Spirit, the Mother. The Spirit is a She. Miryam speaks of the feminine Spirit 167 times in her Gospel. Almost everywhere where the feminine and maternal Spirit is mentioned in Miryam's Gospel, the editors of John's Gospel mention the Father. This conversion must have been completed in a later edition. In the old Hebrew tradition of the First Temple, Hebrew people knew that their divine parents were two: the Divine Mother and the Divine Father. In the second temple period (586 BCE–66 CE) and during an increasing patriarchization process, the Mother becomes lost. But there were secret groups who kept alive her memory and rituals of ascension. There were groups of prophets—called the silent ones—who took refuge abroad, and the Essenes were one group of their successors. We have textual evidence that the Essenes knew of the Father and Mother. In many newfound texts from Nag Hammadi and the Dead Sea you find the Lost Lady, originally a Great Lady. Miryam belongs to this old Hebrew tradition.

2. In Miryam's text she first presents herself as Miryam of Bethany. She was present at Jesus's baptism in the Jordan. They were already together—even before the calling of any other male disciple. The editors of the Fourth Gospel omit Mary's presence at the baptism in the Jordan. In addition, they changed her role from beloved companion to the beloved disciple, a male one. It was not until much later that the name of John was given to this anonymous male beloved disciple. That the beloved disciple would be John is an assumption and not a fact. Several scientists took into account

that the beloved disciple and author of the Fourth Gospel could be a woman, even before the publishing in 2010 of *The Gospel of the Beloved Companion*.[14] Some of them even suggested that this beloved woman could be Mary Magdalene.[15]

3. The role of Mary Magdalene, whom Jesus appointed as Migdalah after the second anointing in Bethany and as his successor at the last meal in Jerusalem, has been written out of the Fourth Gospel, and Peter was pushed forward—Peter, who seems to understand little of higher knowledge.

4. In Miryam's text, there is no conflict between Yeshua and the Judean people as a whole, but between Yeshua and the leaders of the Judean people. Yeshua and Miryam are themselves from Judea and work with love among the common people of Judea, their compatriots. In the Gospel of John and the three Synoptic Gospels—texts that were written later than Miryam's text from the first half of the first century—the original conflict between Yeshua and the Judean leaders of the Temple, the Pharisees, priests, and scribes, developed into a conflict between Yeshua and all the Judeans in an attempt to spare the foreign rulers, the Romans.[16] The role of the Romans was later downplayed to the detriment of the entire nation of Judea. One of the consequences was that a wedge was driven between Christians and Jews. Christians persecuted Jews for "the murder of Christ." It had nothing to do with the original message of love anymore.

5. If you compare Miryam's text with that of John's, it appears that the Fourth Gospel had been changed. Such is the case of the raising of Lazarus, the scene under the cross, and the scene at the open tomb when Miryam sees the risen Jesus. In addition, the editors sometimes contradicted themselves, and there were inconsistencies in the text that had been adjusted to patriarchal

---

14. *MMU*, 540–45 gives the names of these scholars, including James H. Charlesworth, Raymond E. Brown, Ramon Jusino, Esther A. de Boer, and Marvin Meyer.
15. *MMU*, 541–45 gives an overview.
16. *GBC*, 88–89; *MMU*, 86, 90, 211.

standards. Biblical scholars made it clear that numerous clauses were added later. The final chapter 21 was also added in a later edition.[17]

6. Theologians argue that the Fourth Gospel had been edited numerous times and often reshaped.[18] This did not happen with the Gospel of Miryam. It has a rare purity, strength, and coherence. In the second and subsequent redactions of Miryam's source document, which later evolved into the Fourth Gospel of the Beloved Disciple, there are inconsistencies resulting from the fact that the beloved companion's identity and her leadership role had to be concealed.

7. Miryam consistently refers to Jesus as Yeshua the Nazorean. In the canonical Gospels, Jesus was regularly called Jesus of Nazareth, after a very small village in Galilee that was founded around the beginning of the era. This obscured his honorary title and converted it into a place name. We are dealing with Yeshua the Nazorean, the keeper of the covenant, and not with Jesus of Nazareth.

8. Miryam refers to herself as Miryam the Migdalah. Translated, that gives her the name of Mary the Magdalene or Mary the Toweress, the exalted one. In the few places where the name Mary Magdalene occurs in the Fourth Gospel of the Beloved Disciple, later called John, the Greek text reads Mary the Magdalene (Mariam *hè* Magdalene). Here Mary has an honorary title *the Magdalene*.

An example from the Greek text of the Fourth Gospel of the Beloved Disciple, later called John (19:25): *But by the cross of Jesus stood his mother, and his mother's sister, Mary the wife of Clopas, and Mary the Magdalene.* However, the Latin and modern translations feature Mary Magdalene. They changed an

---

17. *MMU*, 553–64.
18. Brown, *The Community of the Beloved Disciple*, 147, 149 suggests three phases in the redaction.

honorary title into a proper name. Now the meaning is lost.[19] The Dutch Willibrord translation of this passage is even worse: it translated her name in this passage as Mary of Magdala. An honorary title is converted into a place name, connected to the place Magdala.[20]

9. Last but not least: in *The Gospel of the Beloved Companion*, Miryam described a sevenfold heavenly journey; her journey to the heavens is more complete than elsewhere. Moreover, in the eighth heaven she meets a luminous and radiant Lady, Spirit. It is this sevenfold path to ascension that is being returned by Miryam to twenty-first century people who are open to it. Ascension is the core theme of this book.

---

19. Nestle-Aland, *Novum Testamentum, Graece et Latine*, the Greek text of John 19:25, p. 313 gives τοῦ ἰησοῦ ἡ μήτηρ αὐτοῦ καὶ ἡ ἀδελφὴ τῆς μητρὸς αὐτοῦ, μαρία ἡ τοῦ κλωπᾶ καὶ μαρία ἡ μαγδαληνή. The Greek gives *Maria hè Magdalènè* or Mary the Magdalene. The Latin translation gives the personal name: Mary Magdalene.
20. The Dutch Willibrord translation gives the name Mary of Magdala; her proper name is linked with the name of the settlement Magdala.

10
THE ASCENSION OF
MIRYAM THE MIGDALAH

The apotheosis of Miryam's Gospel is found at the end of her document. It describes her ascension experience, and her ascension was essential, the essence of her teaching. Miryam's consciousness—ancient texts spoke of her soul—emerged from her body and traveled through the cosmos in subtle bodies. In modern language you could say that she had an out-of-body experience.

Miryam had a vision in which she seemed to leave her physical body. But how do you explain that in modern terms? Researcher Elizabeth Greene wrote a profound article titled "The Celestial Ascent." She stated that the ascension or heavenly journey is scientifically seen as an individual mystical experience. It is about an altered state of consciousness, and the experience is accompanied by specific physiological phenomena. These days the effects of the mystical experiences can be registered by scans—MRIs and other technical means.[1] She concluded that these profound experiences have been transcultural for many millennia. This implied that they are ontological and real.[2] She is right: these visionary experiences occurred in all cultures throughout all times. However, in a secular and postmodern society one hardly finds explanatory models.[3] Modern language (and therefore modern thinking) has no experience

---

1. Greene, 'The Celestial Ascent,' 81n444.
2. Greene, 'The Celestial Ascent,' 81.
3. Greene, 'The Celestial Ascent,' 84

in combining analytical and symbolic thinking,[4] so we can assume, even by modern standards, that Miryam actually experienced her ascension as an out-of-body journey.

Miryam's consciousness rises upon the Tree of Life. From the third and fourth levels, she reached the fifth of nonduality. Finally, from the eighth level, she reached the Kingdom of Light, where she was absorbed in the light. So Miryam went through the whole process of ascension. In addition, she did not die but returned to her physical body at the end of the journey. Afterward, she wrote her Gospel and traveled to France.

What Preceded Miryam's Ascension

In the Gospel of Miryam, Miryam is the first to meet the resurrected Yeshua at the open tomb. It was very early morning the day after the Sabbath. It was still dark, and Miryam had walked the short distance from her home in Bethany to the tomb. When she arrived, she saw that the tomb was open. Inside she saw only empty cloths; the body was gone. She heard a noise behind her and turned away from the entrance of the tomb and saw a man; she did not recognize Yeshua at first, but when he called her name, she recognized him. She wanted to grab ahold of him, but he said, "Miryam, don't hold to me, for I am not of the flesh, yet neither am I one with the Spirit."[5] This important remark shows that Yeshua was still in his ascension process. She did not recognize him in the dark, but he may have looked different. His resurrected physical body may have been more transparent, and bilocation (an ability for a person to be in two places at once) could also have been involved.

After asking her not to hold him, he instructed her to go to his disciples and tell them that she had seen him.[6] She went home and told the good news to the small group of relatives and disciples present there.[7] They all knew that Miryam spoke the truth and were all filled with great joy

---

4. Greene, "The Celestial Ascent," 8.

5. *GBC* [40:6], 75; *MMU*, 459.

6. *GBC* [40:6], 75; *MMU*, 451, 459.

7. *GBC* [40:7–8], 75; *MMU*, 452.

and faith.[8] They decided to send messengers to the other disciples. Miryam mentioned that it would be days before these other people would arrive in Bethany.[9] They fled Jerusalem in fear of the priests and Pharisees. Peter, after what happened in the court of the high priest's palace and being overcome with grief and regret, fled to his region of Galilee.[10] The journey from Galilee back to Jerusalem took three days on foot.

Miryam wrote, and I paraphrase: At the end of that week all the disciples were gathered in the house in Bethany. Now the Migdalah came to them and told them what she had seen and what Yeshua had said.[11] But they were very grieved and wept greatly, asking themselves how to proceed and preach the gospel of the Kingdom of the Son of Humanity.[12] Miryam had shown her leadership by calling everyone into her house. But now she showed it again. She took a prophetic attitude, spoke to them and knew how to turn their hearts to the good. They began to discuss the words of Yeshua.[13]

Peter invited Mary to speak. He called her sister and said that they knew that Yeshua loved her more than any other among women. He asked her to tell them the words of the Rabbi, which she remembered and which she understood, and which they did not know and understand, nor did not hear.[14] So, it is at Peter's invitation that Miryam spoke about higher knowledge or ascension; Peter, who frankly admitted that he and others neither knew nor understood certain words of the Rabbi, while Miryam did.

The parallels between *The Gospel of the Beloved Companion* and the Gospel According to Mary (Magdalene), from the late first and early second centuries, discovered in 1895 in Akhmim in Upper Egypt and not published until 1955, are striking. From certain word usage you can

---

8. *GBC* [40:8], 75; *MMU*, 452, 459.
9. *GBC* [40:9], 75–76; *MMU*, 459.
10. *GBC* [37:7], 70; *MMU*, 407.
11. *GBC* [41:1], 76; *MMU*, 467.
12. *GBC* [41:2], 76; *MMU*, 468.
13. *GBC* [41:4], 76; *MMU*, 469.
14. *GBC* [41:5], 76–77; *MMU*, 470.

deduce that this text was written later than Miryam's text, which dates
from the first half of the first century. Unfortunately, ten of the nineteen pages in the Gospel According to Mary (Magdalene) are missing.
I have discussed the many similarities, but also the differences, in an
earlier book.[15]

Miryam Described Her Own Ascension

After Peter invited her, Miryam described her ascension experience. She
told the assembled group of disciples in the first verse of chapter 42
about the ascension: "What is hidden from you, I will proclaim to you."
And she began to speak to them the words that Yeshua had given her.[16]

In the second verse, she transmitted the words of Yeshua who said
to her: "Miryam, blessed are you who came into being before coming
into being. And whose eyes are set upon the kingdom, who from the
beginning has understood and followed my teachings."[17] Yeshua was
saying here that she understood his teachings from the beginning. This
statement is also found in many other gnostic texts. A Mary Magdalene
tradition emerges here in which Mary Magdalene is described by Yeshua
as the initiate in higher knowledge. The indications that Miryam is the
Woman who has seen the All and that she is the Completion of the
Completions appear in other gnostic texts.[18] That a hidden and lost
Mary Magdalene tradition emerged here—as in other Gnostic Christian
texts—is a great step forward in the Her-Story research.

Yeshua "only from the truth" told her that a great tree is within her
that does not change, summer or winter, and its leaves do not fall; that
whosoever listens to his words and ascends to its crown will not taste
death but will know the truth of eternal life.[19]

---

15. *MMU*, 467–483 and 519–533.

16. *GBC* [42:1], 77; *MMU*, 473

17. *GBC* [42:2], 77; *MMU*, 492.

18. *MMU*, 513-14 with ref. to The Dialogue of the Saviour, NHC III.5 and to Pistis
Sophia, c 19.

19. *GBC* [42:2], 77; *MMU*, 489, 491.

The Tree of Life

In this verse, Yeshua has made it clear that there is a great tree in her, which does not change and there are no leaves that fall. So, it is a symbolic tree. Miryam needed to ascend that tree to the crown, where she will find eternal life. This is a tree inside the human body. You encounter that inner tree in Eastern philosophy, where the journey inward and upward is often experienced internally on a microcosmic level. It involves an inner journey through the seven chakras or wheels of energy.

In verse three, the ascension vision of Miryam continued. She wrote that he showed her a vision in which she saw a great tree that seemed to reach unto the heavens. As she saw these things, he told her that the roots of this tree are in the earth, which is her body. The trunk extends upward through the five regions of humanity to the crown, which is the kingdom of the Spirit.[20] The inner tree inside every human being shows itself in a much larger scale in a great tree outside of every human being. This great tree reaches from the earth to the Kingdom of Spirit in heaven, so the microcosmic tree is at the same time a macrocosmic tree.

What emerges is the triple structure of the roots, the trunk, and the crown.

1. The roots in microcosmic scale are reflected in Miryam's body, where they are to be found in certain areas in and around the Earth. This is where the Lower World emerges, the Lesser Reality of 3D and 4D with the dualistic world of opposites and divisions.
2. Between the roots and the crown, you find the trunk. In the body you will find this area along the spine. In ancient Egypt the spine is erected from the reclining or sleeping position into the standing posture of the resurrected or awakened one. The trunk

---

20. *GBC* [42:3], 77; *MMU*, 489.

Figure 10.1. The microcosmic journey through the chakras. Image of a Shaiva Nath chakra system, folio 2 from the Nath Charit, 1823. Mehrangarh Museum Trust, Jodhpur, Rajasthan, India.

Figure 10.2. Man as a tree, rooted in the earth with the seven chakras and the seven bodies.

also represents the middle region of the soul: the Greater Reality. Above the fourth dimension begins the Middle World of 5D, the area where there is no duality anymore.

3. The crown extends from the eighth level. It is the Upper World or the world of the formless.

People in the East have traditionally been strongly focused on the inner ascension on a microcosmic level from the first root chakra to the seventh crown chakra. In the West people experienced the ascension on a macrocosmic level, where the ascent goes along the seven planets to the fixed stars and beyond.

The doctrine of correspondences is known in both East and West. That means that heavenly bodies influence the people and life on Earth. The seven chakras within a human being are mirrored in the seven levels outside in the cosmos. In the Cretan labyrinth, both Eastern and Western ways are integrated with each other, including: the seven chakras, the seven planetary spheres, the seven notes in a scale, the seven colors in the rainbow, the seven days of the week, and the seven metals. Yeshua said, "The kingdom is without you and within you." He meant that there is a connection between outside and inside.

The many ascension accounts within Western Mysticism described ascents through the planetary spheres, through the ages. However, to my knowledge Miryam was the only woman from classical antiquity to describe an ascension experience. In addition, she is the only one who links ascension to the Tree of Life. Finally, she is personally awaited by a female Spirit, Lady Wisdom in the crown of the tree, which is comparable to the tower room or the holy of holies in the First Temple. Here, you witness how the lost Lady Wisdom, loved and known in Ancient Israel but lost in the reformations of the second temple period after 586 BCE, left her hiding place after 2,600 years of hiding. She showed herself in her full radiant light to Miryam and through Miryam also to us.

The tree is an important symbol for the divine feminine; it is present in the global symbolic system of the Great Mother.[21] The Wisdom Writings of the Old Testament present numerous examples showing that Lady Wisdom or God the Mother is experienced in the tree.[22] A passage in *The Book of Proverbs* gives the irrefutable connection between Wisdom and the Tree of Life:

> She is the Tree of Life to those who lay hold on her.
> Those who hold her fast are called happy.[23]

The Tree of Life as a macrocosmic symbol of connection between Earth and Heaven appears in the Essene scriptures[24] and in the Kabbalah, the much older mystical branch of Judaism where mystical knowledge was often passed down orally. In the Kabbalah, the different levels of the Tree represent the structure of the invisible worlds. It is a kind of plan or map of the heavens. What makes the Kabbalistic tree special is the fact that it consists of a left side and a right side. The feminine and the masculine work together and meet in the trunk in the middle, the third aspect. Only with the ultimate balance between the left and right side do you reach the golden middle way.[25] The Essenes (and later the gnostics) were aware of this principle. It showed itself in the balanced image of God as Father-Mother. It showed itself in the cooperation between women and men in the communities and in the shared leadership of female and male leaders.

There is one more similarity I would like to draw attention to. In the Kabbalah, the lowest level of the tree represents the Earth. It is part of all divine creation. This extends itself from the highest heavens to the sublunary in the lowest realm. There is no trace of the alleged evil of the physical world. It is about Oneness flowing out in Energy, Light, and

---

21. van der Meer, *The Language of MA*, 'The Tree of Life', 405–9; see p. 416–424 for images of holy trees on the art of Old Israel.
22. Sir. 24:13–24 where Wisdom manifests as a tree.
23. Prov. 3:18.
24. Szekely, *The Essene Gospel of Peace*, 106.
25. Baring, *The Dream of the Cosmos*, 41–53.

Figure 10.3. Tree of life based on the Kabbalah, with the right and left sides coming together in the center of the human body, the middle way. Author: Alan James Garner.

Consciousness and manifesting on different vibrational levels. The Essene scriptures showed a great love for the beauty of the Earth, the realm of the Mother, as the teachings of Yeshua and Miryam do. However, both make a difference between the beauty and purity of the Earth and the world of the shadows, where the light only partially penetrates.

The Journey through the Seven Spheres

You can find the Tree of Life in the holy of holies of the First Temple in Jerusalem, in the tower or *debir*. This Hebrew word *debir* is derived from the word *to speak* or *to oracle*; the tower is the place of the oracle and the vision. There it is dark until the lights of the tree are lit.

The tree with the lights is initially round.[26] Later it is pruned to

---

26. Frederick M. Huchel, *The Cosmic Ring Dance of the Angels: An Early Christian Rite of the Temple*, London: The Frithurex Athenaeum, 2016, 57n187 and 188 with reference to the work of Margaret Barker.

Figure 10.4. The macrocosmic journey through the planets. Author: Kwamikagami.

the Tree with the characteristic seven arms. The pruned tree resembles the later menorah, the seven-branched candlestick. The seven torches or oil lamps correspond to the seven celestial bodies, including Moon, Mercury, Venus, Sun, Mars, Jupiter, and Saturn.[27]

The classification by planet spheres is old. The traditional Chaldean order gives from Earth the celestial bodies and planets in this same order. The *Secret Book of John*, a text that is part of the Nag Hammadi library, gives the ascent of the soul through the spheres in the same traditional order. Above the eighth sphere there is the zodiac with the twelve constellations. Above the eighth celestial sphere, with the fixed constellations of the zodiac, is the divine world on the highest level along with the Mother (here called Barbelo) and the Father (or the unknowable

---

27. Barker, *The Gate of Heaven*, 77.

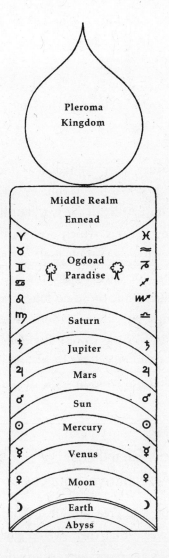

Figure 10.5. Overview of the ascent through the seven spheres to the eighth, ninth, and tenth spheres. From Jacob Slavenburg and Willem Glaudemans, *De Nag Hammadi Geschrift*, 36.

God).[28] The division into seven planetary spheres runs like a common thread through the history of Western esotericism from ancient times, through the Middle Ages to the present day.[29]

The Gospel According to Mary (Magdalene) from Akhmim in Upper Egypt did not describe the heavenly journey along seven planetary

---

28. van den Broek, *Gnostic Religion in Antiquity*, 218; *MMU*, 501.
29. For survey see Greene, "The Celestial Ascent."

spheres, but through four spheres. But that fourth sphere is sevenfold. After that sevenfold fourth sphere, Mary Magdalene reached an area where she was liberated from the world.

Despite the fact that the Gospel According to Mary and the Gospel of Miryam are very similar—some passages are literally the same—there is a big difference.

- In the Gospel According to Mary, Mary Magdalene does not ascend a Tree of Life. She ascends through four spheres with four Guardians before reaching the tranquility of the world of light.
- Miryam ascends a tree with seven gates, which are guarded by seven guardians. Only those guardians guarding levels one to four intend to test the one who passes for purity. From the fifth level there also are gates with guardians, but those no longer test for imperfections or negative feelings with low vibrations. These are no longer there because there is nothing to test anymore. This raises the question of whether those guardians in Miryam's text are really as bad and vicious as encountered in some of the other gnostic texts?

The Watchers or Guardians, the Apocryphal and Alternative Sources

Yeshua continued his teaching about the Tree of Life, recorded by Miryam. He told her in the fourth verse that there were eight great boughs upon this tree, and each bough bears its own fruit, which she must eat in all its fullness. Since the fruit of the tree in the garden caused Adam and Chav'vah (Eve) to fall into darkness, this fruit will grant to her the light of the Spirit that is eternal life. That between each bough there is a gate and a guardian who challenges the unworthy who try to pass.[30]

In later gnostic texts, the guardians are evil. The Dutch specialist in gnosticism Professor Emeritus Gerard Luttikhuizen argued that in the

---

30. *GBC* [42:4], 77; *MMU*, 499–500.

ancient, non-gnostic celestial travelogues, the guardians are not evil, and the descending and ascending souls also receive positive qualities from the cosmic Powers.[31] Another Dutch specialist in gnostic studies, Professor Emeritus Roelof van den Broek, states that ancient philosophers attribute both good and evil qualities to the planetary spheres. In his book with Gilles Quispel, van den Broek discusses Platonists and Hermetists who hold that the soul receives a number of properties from each planet on its descent to earth. In most gnostic writings, the guardians are terrifying and evil, including in the Gospel According to Mary. The guardians want to keep the soul in the lower realms of materiality; they try to prevent the soul from passing through the seven heavens to move on to the eighth, ninth, and sometimes up to the thirteenth heavenly sphere.

Yet there are also gnostic writings that distinguish guardians within the Pleroma or Fullness of Light because the Light World also has a structure of many dimensions where heavenly powers reside.[32] Van den Broek wrote that the lower spheres concern stern planetary guardians, but in the higher spheres they are angels, who are benevolent. In any sphere it remains important to know the passwords or "seals" in order to pass from low to high.[33] Many gnostic texts contain the passwords that one must know.

Miryam wrote that between every branch, there is a gate with a guard who tests those who seek to pass. The guards check whether someone is worthy or "light" enough to continue to the next gate. In fact, they have a positive function, because only people who are worthy are allowed through, as verse 4 shows. Again, you see that Miryam knew an early form of gnosis, which is not yet dualistically colored and

---

31. Luttikhuizen, *De evangeliën van Thomas, Maria Magdalena en Judas*, 126n175 mentions as an example Macrobius, *Commentary on Cicero's The Dream of Scipio* I.12: the descending soul gains reason and understanding in the sphere of Saturn . . . in the sphere of Jupiter power to act. . . . In the sphere of Mars resolution . . . in the sphere of Venus passion.

32. van den Broek, *Gnostic Religion in Antiquity*, 89 mentions "The Untitled Treatise" (Codex Brucianus 2), Zostrianus (NHC VIII.1), Marsanes (NHC X), Allogenes (NHC XI.3), and the Three Steles of Seth (NHC VII.5).

33. van den Broek, *Gnostic Religion in Antiquity*, 89.

offers little information about evil, although it is clearly mentioned. Miryam thinks in Oneness.

In my opinion, Miryam's Oneness philosophy is rooted in Egyptian culture. The emphasis on the Oneness of the created world with the higher worlds is found in the Hermetic writings of the first to third centuries. Their origin lies in Alexandria in Egypt, a melting pot. Three Hermetic writings, which are said to date back to the sage Toth-Hermes, have been found in the Nag Hammadi library.[34] There is a clear connection between the early gnosis and Hermeticism from Egypt, and later gnosticism is seen as radicalization and its naturalization and deformation.[35]

Fixation on Evil in Later Gnostic Writings

I refer once again to the vision of Professor Roelof van den Broek. He stated that "the older Gnostic writings are strongly interested in the origins of ignorance and death and the question of how man can be delivered from them. . . ."[36] In the later Gnostic texts of the second and third centuries, naturalization and fragmentation occur.[37] They lose sight of the Oneness of the created world in their many attempts to explain the evils of the world. They become fixated on evil and the many realms of darkness, which are filled with many fallen angels."

Only later does radicalization set in, and gnostic texts refer to the fall of Sophia and her birthing the evil creator god. This creator god is

---

34. The Discourse on the Eighth and Ninth (NHC VI.6), The Prayer of Thanksgiving (NHC VI.7), and the Asclepius (NHC VI.8).
35. van den Broek and Quispel, *Hermetische Geschriften*, 83; van den Broek and G. Quispel, *Corpus Hermeticum*, 22–23.
36. van den Broek, *Gnostic Religion in Antiquity*, 26, 82.
37. van den Broek, *Gnosis in de Oudheid*, 48 speaks of "radical gnosis" or "mythological gnosis" of the second and third centuries; van den Broek, *Gnosis in de Oudheid*, 90 is about the books of Jeou "usually dismissed as products of a feral gnosis and sometimes dated to the first half of the 4th century"; van den Broek, *De taal van de Gnosis*, 14; Luttikhuizen, *De Evangeliën van Thomas, Maria Magdalena en Judas*, 82: the Gospel of Thomas knows no Fall.

also called the "Demiurge" in the gnosis. The evil Demiurge is identified with the cruel God of the Old Testament. Van den Broek called
the origin of the evil Demiurge and his identification with the cruel
God Yahweh of the Old Testament the most difficult problem in gnosis
research.[38] This problem is solved once you understand that the First
Temple tradition was suppressed by the second. In the eyes of the refugees who remained faithful to the old First Temple traditions in the
diaspora, the rulers in the second temple and their second God Yahweh
were the incarnations of the fallen angels. That was the origin of the
conflict between the Pharisees and Yeshua.

That evil creator god had helpers. They were called Planetary
Powers, or Powers, in later gnostic texts. The name Archons or World
Rulers also occurred. Another name is the Watchers. This is about
angels watching over humanity. The Watchers watch and guard.[39]
They guard the seven gates. The soul must speak the correct passwords
before the guardians will let it through.

The First Book of Enoch contains the Book of the Watchers.[40] Here
you will find the stories about two different groups of fallen angels. The
First Book of Enoch has ended up in the extra-biblical books under the
heading of pseudo-epigraphic.[41] Yeshua and his followers had considered it canonical and quoted from it. Some of those Watchers, contrary to divine command, coveted the fair daughters of men. Enoch
stated that two hundred Watchers descended from heaven on Mount
Hermon. After sexual intercourse with the earthly women, this group
of guardians begat evil giants from the earthly women, and these giants
caused much mischief on Earth. The guardians who landed (or were
stranded) on Earth, together with other groups of fallen angels, also
taught people about astrology, spells, medicine, the processing of metals

---

38. van den Broek, *De Taal van de Gnosis. Gnostische teksten uit Nag Hammadi*, 172.
39. The Discourse on the Eighth and Ninth, NHC VI.6.62.
40. 1 Enoch 1–34; see R. H. Charles, *The Book of Enoch*, 31–55; Barker, *The Lost Prophet*, 22 prefers another division.
41. Charlesworth, *The Old Testament Pseudoepigrapha. Volume 1*, 3–89; Barker, *The Lost Prophet*, 18.

to use for weapons, and many other heavenly secrets, which they should have not disclosed to humanity.[42] The spirits of the evil giants remained on Earth and incarnated in humans.[43]

This would be the main reason for the degeneration of humankind. However, synagogues and the Church blame the sin of Adam and Eve as the personification of all humanity—that human beings are guilty of causing evil. But the cosmic evil came first. Where else did the serpent that tempted Eve (according to the myth added later in Genesis) come from?[44]

Genesis 6 gives a much abbreviated and more positive version of the events covered by Enoch, describing how rebellious sons of God begat evil giants with the daughters of men. In Genesis they are presented as mighty men that were of old, the men of renown.[45] The editors of the Old Testament aimed to cover up the traces of fallen angels.[46] Yet the fallen angels were not forgotten: you find the books of Enoch in Qumran texts; you find quotes from Enoch with Yeshua and Miryam and their followers; you will find elements of the Enoch tradition in a number of gnostic texts from the Nag Hammadi library.[47]

Several passages in The First Book of Enoch mention angels or Watchers who have not fallen. They continue to watch closely from the heavens, in contrast to the two hundred princes, the former guardians, who fell from heaven. The heavenly Watchers are also called the Holy Ones, something that is repeated in the book of Daniel.[48] So it is not right to describe all the Sons of God, the Holy Ones, as evil. There are

---

42. 1 Enoch 7.1, see Charles, *The Book of Enoch*, 35; Barker, *The Lost Prophet*, 39.

43. 1 Enoch 15:8–16:1, see Charles, *The Book of Enoch*, 43.

44. Barker, *The Lost Prophet*, 23, 33, 38.

45. Gen. 6:1–6 weaves two traditions of fallen angels into one pattern, confuses these, and turns the negative into the positive; van der Toorn, Becking, and van der Horst, *Dictionary of Deities and Demons*, "Watcher," 1681–85, 1681.

46. Barker, *The Older Testament*, 114; Barker, *The Great Lady*, 331: "The development of the Targums shows that the fallen angels were gradually written out of the tradition, but not exactly when this happened."

47. The Secret Book of John, just before the Descent of Pronoia, NHC II.1.75–77; Barker, *The Lost Prophet*, 112: many Enochian ideas return in later gnostic texts.

48. Daniel 4; van der Toorn, Becking, and van der Horst, *Dictionary of Deities and Demons*, "Watcher," 1681–85, 1683.

different groups discoverable through reading the Gospel of Miryam.

I now make a small detour to the alternative sources. Daniel Benezra also distinguished between the heavenly and the fallen Watchers. He was able to project his consciousness into the future. Then, at the end of planet Earth's cycle, he saw a greater understanding emerge between humans and angels. He saw them working together more: "The guardians will come into closer contact with us."[49] He clearly means here the heavenly Watchers, the Holy Ones, so here you also find a positive view of the heavenly Watchers.

To Prentis's question about whether the guardians are beings from another solar system, who have come to observe human life, the surprising answer is: "Yes, they come from other star systems and observe human progress or rather the lack of it. They have a deep understanding of the Divine Plan for the entire universe; they see our little dramas on planet Earth in a much broader context."[50]

The Watchers are thus angels and beings who come from beyond the Earth and who can be divided into two groups: the heavenly Watchers and the fallen Watchers who have lost their light. Much is known about both groups in the First Temple period.[51] Afterward the traces of the fallen angels are erased in the canonical books of the Old Testament. Despite the fact that the followers of Yeshua and Miryam are aware of the older traditions about the fallen angels of the First Temple, later church fathers of the Christian tradition whitewashed dark forces again by writing them out of history.[52] With the finding of the books of Enoch, the Nag Hammadi library, and the Dead Sea Scrolls, this knowledge is now returning to us.

---

49. Wilson and Prentis, *The Essenes*, 248.
50. Wilson and Prentis, *The Essenes*, 248.
51. Barker, *The Lost Prophet*, 94: 1 Enoch origins in the First Temple tradition; 107: it is a composited book whose fragments are older than the third century BCE.
52. Elizabeth Clare Prophet, *Fallen Angels and the Origin of Evil: Why Church Fathers Suppressed the Book of Enoch and its Startling Revelations*, Missoula, MT: Summit University Press, 2000, 7. The church fathers knew that fallen angels could incarnate. From the fourth century they considered this knowledge so dangerous that it was proclaimed heretical.

In verse 5, Yeshua continued telling Miryam that the leaves at the bottom of the tree are thick and plentiful so no light penetrates to illuminate the way . . . "But fear not, for I Am the way and the light, and I tell you that, as one ascends the tree, the leaves that block one from the light are fewer, so it is possible to see all more clearly. Those who seek to ascend must free themselves of the world. If you do not free yourself from the world, you will die in the darkness that is the root of the tree. But if you free yourself, you will rise and reach the light that is the eternal life of the Spirit."[53]

Miryam made it clear that it is very dark at the roots of the tree because no light penetrates. The darkness at the roots of the tree represented the world, and that world is full of darkness, but as you ascend it gets lighter. For this reason, you must free yourself from the world. In several places in Miryam's Gospel, such as here, the word *world* has an extremely negative connotation. You must free yourself from the world, otherwise you will die in darkness at the root of the tree. In Miryam's document, Yeshua regularly talked about the world, saying to the people: "The world cannot hate you, but it hates me, because I testify about it that its works are evil."[54] The world is dark; not that the Earth and matter are bad, but the works of the world are bad. The evil of the world has to do with cosmic evil: the presence of the fallen angels, the beings from beyond the Earth and their progeny. That has brought darkness.

The Seven Fruits of the Tree of Life

Miryam recounted in verse 6 that after he had explained these things, she felt her soul ascend and she saw the first great bough that bore the fruit of love and compassion, the foundation of all things. She knew that before you could eat of this fruit and gain its nourishment, you must be free of all judgment and wrath. Only when you have freed yourself of these burdens may you eat of the fruit and so gain the love

---

53. *GBC* [42:5], 78; *MMU*, 502–03.
54. *GBC* [20:2], 38; compare Jn. 7:7; *MMU*, 225

and compassion that will allow you to pass the first of seven guardians. She wrote that she heard the voice of the Lord of Wrath calling her but that she denied him, and he had no part in her.[55]

So at the first gate, the watchful Watcher, the Lord of Wrath, checked whether Mary's soul had rid itself of the burden of judgment and wrath. Not judging, really being detached, and staying above contending factions in all circumstances is difficult. In fact, you are already living in a state of nonduality. Miryam appeared to have freed herself from judgment and anger. The Lord of Wrath had no hold over her; he could no longer engage with her because those feelings of low vibration were no longer alive in her. Miryam could enjoy the fruit of love and compassion, after which there was an important remark: love and compassion are the foundation of all things. That is quite something: the cosmos is thus built on love and compassion of high vibration. Those two feelings connect you to the web of higher dimensions, and through those connections you are part of the whole web of life.

In verse 7 Miryam said that she saw her soul ascend again and that he showed her the second great bough, weighed down with the fruit of wisdom and understanding. She understood that before you can taste of its bounty, you must be free of all ignorance and intolerance. Only then can you eat of the fruit and so pass upward unhindered through the second of the seven gates. She heard the voice of ignorance call to her, but she knew him not, and so her soul was thus unchallenged.[56] Here, she discarded ignorance and intolerance. Taking off the veils of ignorance required trying to gain understanding, to wake up and stand up, to first acquire the knowledge that gives insight into things and learn to open your heart. Intolerance had no place here, whereas tolerance did because everything is connected in that web of love and compassion.

In the Gospel According to Mary from Akhmim of Upper Egypt, several pages are missing. In this text, after a gap, the ascent begins from the second level and continues through the fourth level. Here the

---

55. *GBC* [42.6], 78; *MMU*, 503–04.
56. *GBC* [42:7], 78; *MMU*, 504–05.

evil Watcher of the second sphere is called Desire. The text reads on sheet 15.1: "And Desire said: 'I did not see you go down, but now I see you go up. Why do you lie? You belong to me!' The soul replied: 'I saw you. You did not see me and did not know me. I was a piece of clothing to you, and you did not know me.'" When she said this, she went away with great joy.[57] Here, you see that the soul has descended and clothed itself in various garments or subtle bodies. After the descent follows the ascent and the removal of those garments. The guardian Desire tests the soul. But it no longer has desire in it and has already ascended without Desire seeing or recognizing it. At a higher level you see lower levels, but not the other way around.

In verse 8 Miryam continued telling the disciples that her master showed her the third great bough, which bore the fruit of honor and humility and that only when you are free of all duplicity and arrogance may you partake of its nourishment. She said that arrogance called to her, saying that she was not worthy and had to go back. But her soul was deaf to him, and so she moved onward and upward into increasing light.[58] Miryam seems to be saying that you cannot serve two masters; you cannot serve light and darkness. Don't be duplicitous: walk your talk, be honest, and be a person who is not duplicitous. Arrogance comes from an inflated ego, from the little self, which is disconnected from the web of life and lives in separation and duality. Arrogance has a loud voice and wants to attract attention by venting irritation. This no longer suits a person when the heart is purified. There is a change to an attitude to modesty that deserves praise and honor.

Miryam continued her ascent in verse 9 by telling the disciples that there came the fourth bough, blossoming with the fruit of strength and courage. She heard Yeshua tell her that to eat of this fruit, you must be free from the weakness of the flesh and have confronted and conquered

57. Meyer, *The Nag Hammadi Scriptures*, "The Gospel of Mary," 743–44; *MMU*, 504–05.
58. *GBC* [42:8], 78–79; *MMU*, 505–06, 508.

the illusion of your fears.[59] The Master of the World stood before her and claimed her as his own, but she denied him, and he had no part in her.[60]

Discarding the weakness of the flesh means to be free from the temptations of the world. It means not being distracted by the many things that attract attention in the outside world, such as fleeing into temptations, engaging in addictions, and placing too much attention on material things. It is tested whenever you are still afraid of anything: of loss, of failures in life, illness, old age, or death in any form. Can you accept life as it comes and surrender and trust? Can you eventually see the positive in negative events? This works when a connection with the web of life is woven. You succeed when you know yourself to be a link in the chain and your confidence grows that everything is good for something, even if you don't know why. Those invisible hands help when someone on Earth calls it, and you may enjoy the fruit of strength and courage—for it takes strength and courage to face the Master of the World, the Trickster, and leave him behind.

Miryam continued the teaching in verse 10 and told the gathered disciples that her master had told her that when you have rejected the deceiver, you can pass through the hardest gate of all to attain the fifth bough and the fruit of clarity and truth; that you will know yourself for the first time and understand that you are a child of the living Spirit. As her soul moved upward, she realized that she could no longer hear the voice of the world, as all had become as silence.[61] In verse 10 the Master of the World from verse 9 is called the Deceiver. Only when you leave him behind do you pass through the most difficult gate of all seven gates: the gate from the fourth to the fifth branch or dimension.

Here yawns the huge gulf between the worlds of the third and fourth dimensions on the one hand and those of the fifth on the other.

---

59. *GBC* [42:9], 79; *MMU*, 506 and 508.
60. *GBC* [42:9], 79; *MMU*, 508.
61. *GBC* [42:10], 79; *MMU*, 508–09.

They are separated from each other by a veil that casts a shadow in the lower world: the shadow world. It is dark and gloomy there, especially at the root of the Tree of Life. And right at that transition from 4D to 5D is the ultimate superstructure of all evil: the Master of the World. Between the fourth and fifth levels, the ultimate confrontation with the Master of the World takes place. The title "Master of the World" is used more often in Miryam's Gospel.[62] There are also designations such as the Ruler of the World,[63] the Prince of the World,[64] and the King of the World.[65]

The world of darkness has a ruler, or a Prince of the World. Earlier, the two hundred princes or Watchers who descended were mentioned. Satan, as a Son of God, was originally part of the heavenly royal court of princes. Prior to his fall he was one of the Elohim.[66] Earlier in Miryam's gospel (chapter 25), Yeshua said to the Pharisees "Why do you not understand my speech? Because you cannot hear my Word. You are of your father, the Prince of the World, and you want to do the desires of the world."[67]

And immediately after this Yeshua said, "He was a murderer from the beginning and does not stand in the truth, because there is no truth

62. *GBC* [6:8], 19; *MMU*, 112, 119: The disciples wondered at what he had done, but he said to them, "Why do you marvel thus? Have I not told you that I am in the Spirit as the Spirit is in me? It is man who sees only poverty, for he sees with the eyes of the master of the world. But where man sees poverty, the Spirit sees only abundance. What the Spirit sees, I see, and what I see, the Spirit sees. And what the Spirit sees is."; *GBC*, [42:9], 79; *MMU*, 506–08.

63. *GBC* [25:3], 44; *MMU*, 276: And Yeshua said to them, "You are of the world. I am of the Spirit. If you choose this world, then you are the bondservant of its rulers. I am not of this world, and the ruler of this world has nothing in me. Tell me; if you are of the world and your world ends, where then will you go? Only from the truth I tell you that unless you believe that I am sent by the Spirit, you will die in your darkness."; *GBC*, [30:5], 57; *MMU*, 328–29.

64. *GBC* [25:12], 46; *MMU*, 276; *GBC*, [28:6], 53; *MMU*, 311.

65. *GBC* [35:23], 67; *MMU*, 400.

66. Freedman, Myers, and Beck, *Eerdmans Dictionary*, 1169: Satan was one of the sons of the Elohim.

67. *GBC* [25:12], 46; *MMU*, 273.

in him. When he speaks a lie, he speaks on his own, for he is a liar, and the father of it."[68] Finally, in verse 6 of chapter 28 there again the Prince of the World is mentioned and Yeshua warned his hearers: "Be on your guard against the Prince of the World, for he is darkness and the father of it, and would have you live in the bondage of your own fear."[69]

It is tempting to go deeper into this phenomenon that is called the Prince of the World. Where does the ultimate evil come from? For many centuries it was hidden away but is known again. For now, it is enough to know that it is there. Because at the moment of sufficient light frequency, Miryam leaves it behind and it disappears. What she teaches us is to focus on the light and ignore the dark so it won't grow. Pay attention to the light and you grow in light.

From this moment on, Miryam, in passing the higher regions, meets no negativity anymore, although there were guardians. They are more like gatekeepers signaling the transition to a new and higher frequency of light. From the fifth dimension, the soul begins to heal from the many traumas it has suffered.

In verse 11 of chapter 42, Miryam continued her ascent and described that she saw in the light above, the sixth bough, the one that bore the fruit of power and healing. Her master told her that only when you truly have eaten the fruit of clarity and truth of oneself can you partake of the fruit of power and healing: the power to heal your own soul and thereby make it ready to ascend to the seventh bough, where it will be filled by the fruits of light and goodness.[70]

In verse 12, she stated that she saw her soul, free of all darkness, ascend again to be filled with the light and goodness that is the Spirit, and she was filled with a fierce joy as her soul turned to fire and flew upward in the flames from whence her master showed her this eighth and final bough, upon which burned the fruit of the grace and beauty of the Spirit.[71]

---

68. *GBC* [25:13], 46; *MMU*, 273.
69. *GBC* [28:6], 53; *MMU*, 300.
70. *GBC* [42:11], 79; *MMU*, 509.
71. *GBC* [42:12], 79–80; *MMU*, 510.

Behind the Eighth Portal:
The Lady

Verse 13 follows the apotheosis. Miryam wrote: "And I felt my soul and all that I could see dissolve and vanish in a brilliant light, in a likeness unto the sun. And in the light, I beheld a woman of extraordinary beauty, clothed in garments of brilliant white. The figure extended its arms, and I felt my soul drawn into its embrace, and in that moment, I was freed from the world, and I realized that the fetter of forgetfulness was temporary. From now on, I shall rest through the course of the time of the age in silence. And then, as if from a great distance, I heard the voice of my master tell me, "Miryam, whom I have called the Migdalah, now you have seen the All, and have known the truth of yourself, the truth that is I Am. Now you have become the completion of the completions." And thus, the vision ended.[72]

In no ascension account is there a Lady who waits for someone at the end of the process and joyfully embraces him or her. Until Miryam wrote her ascension record, in religious history only men made the heavenly journey, and none of them met a Lady at the end of the journey. The soul of Paul, for example, encountered a world creator, an old man, in the seventh heaven. It is crystal clear that Miryam has been a role model and example, especially for women in antiquity. Who is the Lady who waits for her? In the Hebrew tradition it was about Spirit or Lady Wisdom. Spirit was mentioned by Miryam 167 times in her text; she had not been changed yet into the Father, as in the Gospel of John. Miryam had reached the bridal chamber, and the sacred marriage between the lower and the higher self had been consummated.[73]

In Miryam's account the Lady had no name. From the First Temple symbolism we know it is the Mother, Lady Wisdom. In gnostic texts you also come across the name Barbelo. She was the central figure in the

---

72. *GBC* [42:13], 80; *MMU*, 511–14.
73. van den Broek, *Gnostic Religion in Antiquity*, 104 is about the union in the heavenly bridal chamber.

Figure 10.6. Mary in the top of the illuminated and fiery tree of life or menorah, from a Byzantine manuscript: *The Physiologus*, Byzantium, eighth century.

Pleroma. Barbelo was regarded as the first manifestation of God's inner thoughts. She was therefore regarded as the principle of all knowledge and salvation. In some texts she divided herself into three lights: eons, spaces, or dimensions, and ascension goes through these three levels.[74]

The bridal chamber can be found in the First Temple in the higher holy of holies, in the tower or debir. I repeat: the Hebrew word *debir* is derived from the word *to speak* or *to oracle*; the tower is the place of the oracle and the vision.[75] There it is dark until the lights of the Tree of Life are lit. The tree gives light, and that light is reflected a thousandfold from the golden walls with which the holy of holies in the tower is lined.[76] That makes it very hot there, which is why the people who undergo such a heavenly journey wear radiant white linen clothing. It is

---

74. Luttikhuizen, "Monism and Dualism in Jewish-Mystical and Gnostic ascent texts," 749–775, Marsanes and Allogenes reached the highest level; Zostrianus, NHC VIII.29 17–20 did not.

75. Barker, *The Revelation of Jesus Christ*, 13.

76. Regarding the almond tree pruned into seven arms, the Greek and Latin word for "almond" is *amygdala*; the Hebrew word for almond means "Great Mother."

the clothing of the angels and the holy ones, men and women. Wisdom, at the end of Miryam's vision, wore the brilliant white garments worn by her children in her tower chamber, the bridal chamber. The Tree of Life gives light, as the knowledge of Wisdom brings enlightenment. Whoever is initiated by Wisdom into her mysteries becomes a consecrated one—an angel, a saint, a holy one, an awakened one—awakened in the light body.

Miryam closed the chapter with two verses, 14 and 15. She told the disciples: "This is what my master has told me and has shown me. And only from the truth, I tell you, that everything I have revealed to you is true."[77] Her closing words were: "When the Migdalah had related all that Yeshua had said and done, she became silent because it was in that silence that Yeshua had spoken to her, and revealed these truths."[78] The squabbles of the world have ceased; there is peace and quiet. In gnostic circles, rest is the indication for the state of happiness that arises upon the return to the light world.[79] Here the vision ends. Miryam ends her out-of-body journey and lands in her physical body again.

For me, Miryam's description of her ascension is the pinnacle of her text. I have compared her ascension to other ascension accounts, and I stand by this. I am convinced that her heavenly journey is of great significance for the people of today. She is a Wayshower. In her teaching, ascension is the essence, and therefore ascension is the main theme of this book.

---

77. *GBC* [42:14], 80; *MMU*, 514.
78. *GBC* [42:15], 80; *MMU*, 514.
79. Luttikhuizen, *Gnostische Geschriften* I, 58.

11

ASCENSION
AS A PROCESS,
PAST AND PRESENT

In this chapter we follow the descent of Spirit through three realms. On the return journey it goes through these three areas in five steps. Ascension is a process of taking more and more light into the energy bodies. That was the case in the past, and it is the case now. Yet there are important differences in the ascension process from the past and now. More information about this topic will follow at the end of this chapter.

The Three Realms:
The Apocryphal Sources

Spirit descends through three realms.

3. The higher world or spiritual world of Spirit.
2. The middle world or world of the soul.
1. The lower world or physical world of the body.

You encounter the tripartite division in *The Gospel of the Beloved Companion*. The Tree of Life through which Miryam ascends has three levels.

Figure 11.1. Holy Spirit Overshadowing Marian Monogram.
Stained glass panel from Our Lady of the Rosary Shrine at
Saint Patrick Church in Merlin, Ontario.

1. The roots represent the earthly level and the body.
2. The trunk represents the middle world and is divided into five areas.
3. The crown stands for the Kingdom of Spirit.[1]

You can also read about the three areas in *The Three Steles* (Pillars) *of Seth*, a text from the Nag Hammadi library from 1945. It is a writing of the first Christians who practiced the Way as a process and were part of the inner group or the mystery school. They felt strongly connected with the Spirit. While the first hymn in *The Three Pillars of Seth* was traditionally seen as a hymn to the Father, John D. Turner saw it differently in his revised and updated edition of the Nag Hammadi texts. He also identified the first hymn as a hymn to the Mother.[2]

---

1. *GBC* [42:3], 77; *MMU*, 497.
2. Meyer, *Nag Hammadi Scriptures*, "The Three Steles of Seth," NHC, VII.5.120.17–121.16, 526–36, 528, with reference to the Secret Book of John.

The conventional translation is therefore based on the hymn to the Father.

> *He, who caused the maleness*
> *that really are to become male three times,*
> *He who was divided into the pentad,*
> *He who was given to us in triple power,*
> *He who was begotten without begetting,*
> *He who came forth from which is elect,*
> *because of what is humble he went forth from the midst.*[3]

Margaret Barker, based on Turner's new translation, translated something completely different. Suddenly a radiant and luminous Lady came forward, a Lady who waited for Miryam in the eighth sphere of heaven. Barker's updated translation of the Mother, Spirit, or Wisdom reads:

> *You who made the shining ones,*
> *You who truly are shining threefold,*
> *You were divided into five,*
> *and given to us in threefold power,*
> *You were generated with birth.*
> *You came from the upper state,*
> *and entered the middle*
> *for the sake of the lower.*[4]

What concerns me in this passage are the three worlds through which Spirit crosses.

---

3. Text according to the conventional and older English translation of Robinson, *The Nag Hammadi Library*, 398; see also the Dutch transl. in *De Nag Hammadi Geschriften*, 855.
4. Barker, *The Great Lady*, 22. This full translation of the English text is based on the updated text of John D. Turner of The Three Pillars of Seth, NHC VII.5.120, and was given by Barker during the summer school, on Thursday, 10 August 2022.

3. You came from the upper level (heaven);
2. And entered the middle one;
1. For the sake of the lower.

The feminine Spirit is a threefold force: "You who truly are shining threefold." But she is also fivefold: "You were divided into five."

According to Turner and Barker, this is not about a He, but about a She. According to Barker this passage is notorious for the fact that the Coptic translator spoke crippled and deficient Hebrew, and in the Coptic translation of Hebrew, he translated it patriarchally. In this passage Spirit is invoked in a sung hymn. Spirit here stands for Wisdom, the Mother or Barbelo. Barker's translation differed greatly from conventional translations. She started from an original Hebrew text and argued that the Coptic translator did not have a good command of the original Hebrew and made mistakes.[5] She put forward all kinds of arguments about why the Coptic texts in the Nag Hammadi library were originally based on Hebrew texts.

Miryam the Migdalah ascended along three planes: from the root through the trunk to the crown. In numerous Gnostic Christian sources you read that Spirit descends through three realms and guides the process of ascent through three realms. What do the alternative sources say about this?

The Three Realities: The Alternative Sources

Alariel did not speak of realms or regions or spheres or levels; he spoke of realities, mentioning three.

3. The Upper Reality
2. The Greater Reality.
1. The Lesser Reality.[6]

---

5. Barker, *The Great Lady*, 22.
6. Wilson and Prentis, *Beyond Limitations*, 99.

The Upper Reality consists of the Unmanifest or Nonmanifest; it is above or inside the manifested universe. Wisdom teachers of the Native American peoples speak of The Great Mystery. It is the Divine, beyond space and time, where a blueprint of the divine plan is kept, where there is cosmic silence from which all creation flows. It is also called the Source or the Void; it is outside of space and time. It is the unmanifest seed from which flows the manifest creations, which create time and space. Eventually, they will flow back into that again. The void is without limit, without time and without space. It is.

The Greater Reality starts with the fifth dimension at the lowest level and goes further up; this is where the reality of nonduality begins.

The Lesser Reality is daily life on planet Earth, essentially an experience in the third dimension or 3D. After discarding the physical body, you reach 4D. You will experience a period of rest, recovery, and reflection. It is also called the Interlife or the in-between area: the fourth dimension or 4D. The 3D and 4D worlds form a closed loop of experience. This has been described as the wheel of birth, death, and rebirth. When you die, you enter 4D and reincarnate many times in 3D. However, you must gather more luminosity to ascend further and enter 5D.[7]

Miryam describes in her gospel the Tree of Life with three areas: the roots, the trunk and the crown. With this explanation of Alariel's we can make the connection to the three levels of the tree, which in modern terms are the three main dimensions.

1. The roots represent the physical and subtle world of 3D and 4D.
2. The trunk represents the middle world from 5D and on.
3. The crown represents the unmanifested, with Miryam, from 8D.

In certain heavenly journeys, it goes up to 13D before you reach the formless realm.

Miryam ascended to 3D and 4D. She came to the great chasm that separated her from 5D, and it is precisely in this upper level of 4D that

_____

7. Wilson and Prentis, *Beyond Limitations*, 99.

she encountered the pinnacle of evil, the Master of the World. The Dutch author Marcel Messing states that fallen angels reside in 4D. This division into 3D and 4D on the one hand and 5D and higher on the other, is found in Miryam's Gospel. Her writing points the way to nonduality and higher dimensions. She explained what ascending the Tree of Life essentially means.

Alariel argued that he disagreed with spiritual teachers who believed that the human experience in 3D is an illusion. What is an illusion is to think that this level of reality is the only reality in the universe. Alariel said: "It is important to recognize and respect all levels of creation."[8]

Prentis asked him "Why is the Lesser Reality imperfect because surely a perfect God can realize a perfect creation?"

To which Alariel replied: "The Lesser Reality is indeed imperfect because it is not connected to Truth and Oneness, but it is perfectly adapted to your needs at the 3D level of experinece. In the wisdom of God, sometimes even the imperfect can be perfect.[9] From 3D, the higher dimensions cannot be perceived with the normal senses. For example, those who have died and are present in the fourth dimensional reality—also called the Interlife or the space between lives—will not be visible to anyone living in 3D, who has the ordinary vision of that plane of existence. In the same way fifth-dimensional beings will not be visible to those in the Interlife or the in-between realm, who possess only 4D sight. The Greater Reality from 5D on contains numerous other dimensional realities."

When asked by Prentis how many dimensions the Greater Reality encompassed, Alariel refused to answer.

It would distract from the task facing humanity today: to transcend the 3D paradigm with all its obsession with duality and step into the non-dual existence of the fifth dimension. The Greater Reality is founded on and reflects an essential Oneness, and the distance from the Lesser to the Greater Reality is moving towards and merging with

---

8. Wilson and Prentis, *Beyond Limitations*, 100.
9. Wilson and Prentis, *Beyond Limitations*, 100.

that Oneness. Oneness will require a complete reorientation of your consciousness, and that will be a major shift for you... Vengeance, for example, becomes absurd within the frame of Oneness. If all life is an interconnected web of being and consciousness, then who exactly would you be taking revenge upon? Your aim should now be to focus on the Oneness that you all share, rather than any small differences that may seem to divide you. You will be challenged to stay in that truth, whatever happens in your life. This is when the universe will test you and events will occur to plumb the depth of your commitment to the truth. Living in non-duality and unity and not judging and forgiving can be extremely hard and intense. It is difficult not to hate but to love your enemies; you can condemn their actions, but not condemn them as beings. The challenge now facing humanity is to lay aside the divisive dualistic way of thinking and feeling and start looking for common ground, the basis for friendship and cooperation.[10]

The Five Steps on the Path: The Apocryphal Sources

In various apocryphal texts the five seals of the Mother are mentioned:[11] Protennoia or foreknowledge, also called Barbelo, states in *The Three Forms of First Thought*, or *Trimorphic Protennoia*, a Nag Hammadi text:

(49) One who possesses the Five Seals of these names has stripped off the garments of ignorance and put on shining light...

(50) I proclaimed the ineffable Five Seals to them, so that I might abide in them, and they also might abide in me.[12]

---

10. Wilson and Prentis, *Beyond Limitations*, 101–03.
11. The Acts of Thomas chapters 27, 49, and 132; Klijn, *Apocriefe Handelingen van de Apostelen*, 116; Klijn, *Apocriefen van het Nieuwe Testament*, II, 82, 97.
12. Trimorphic Protennoia, NHC XIII.1.49 and 50; Turner, "Three Forms of First Thought" in Meyer, *The Nag Hammadi Scriptures*, 715–35; Robinson, *The Nag Hammadi Library*, 521.

The five seals belong to the mystery knowledge. In the Gospel of Philip—a text that offers a continuation of the legacy of the First Temple, Philip states: "The Lord did all in mystery: a Baptism and an Anointing and a Eucharist and a Redemption and a Bridal Chamber."[13]

Those mysteries of Gnostic Christians are accomplished on three levels of Spirit activity. They form a kind of three-stage rocket to the light and can be compared to the three ascending rooms in the temple. Philip in his Gospel linked the ascent through the three chambers of the temple with the five mysteries associated with it. In logion 69 he provided the three rooms of the temple:[14]

1. The holy on the first and lowest level
2. The holy of the holy, higher up, on the second level
3. Holy of holies on the top and third level

It is not about a physical temple but about a nonphysical, spiritual temple that you build in your heart. The ascent through the three temple rooms is compared to the ascent of Wisdom through her three realms:

1. The Lower Kingdom or physical world
2. The Middle Kingdom of the soul
3. The Upper Kingdom of Heaven or the Spiritual World

Thereafter, in the Gospel of Philip, in the same logion 69, the five seals or mysteries are divided into three frequency levels or realms:

1. The Baptism, The Anointing, and The Eucharist are received on the lower level.

---

13. Gospel of Philip, NHC II.3. logion 67, in Meyer, *The Nag Hammadi Scriptures*, 150; Luttikhuizen, *Gnostische Geschriften*, I, 95; *De Nag Hammadi Geschriften*, 338; *MMU*, 36, 41, 58, 84, 247, 333, 336, 501, 509, 511.
14. Gospel of Philip, NHC, II.3.69, in Meyer, *The Nag Hammadi Scriptures*, 175; Robinson, *The Nag Hammadi Library*, 151; *De Nag Hammadi Geschriften*, p. 340; *MMU*, 326.

2. Redemption on the second level; it is about breaking free from the material world through Crucifixion and Resurrection.

3. The Oneness in the bridal chamber is experienced on the third and highest level.

In modern terms it is about ascension.[15] It is this fivefold path to the Light that we are going to follow in Miryam's Gospel.

The First Step:
The Baptism or the Second Birth

In the Gospel of Miryam you read that Yeshua the Nazorean and Miryam of Bethany arrived at the Jordan River where it flows into the Dead Sea—the place where John baptized people. Miryam and Yeshua were together, which indicates the couple were officially married based on the expectations of their time. John the Baptist had a vision in which he saw the Spirit descending from heaven like a dove. He saw how She continued to rest on him. He heard Her say, "The one on whom you see the Spirit descending and resting will baptize in the Spirit."[16]

Elsewhere I have explained, based on texts from Hebrew Christianity, that here resounds a woman's voice from heaven: the Voice of the Mother.[17] The Gospel of Miryam synchronizes with these ancient sources where the Spirit is a She. And one more thing: I believe that She or Spirit descends in the form of a dove on both Yeshua and Miryam: Both go through a baptismal experience, and both saw the heavens open.

Yeshua and Miryam saw the heavens open and had a vision. Prophets and visions were part of the ancient Hebrew First Temple tradition, before second temple Judaism forbade it. You were reborn in Light; you

---

15. Gospel of Philip, NHC, II.3.69, in Meyer, *The Nag Hammadi Scriptures*, 175: "The holy place is baptism, the holy of holy is redemption; the holy of holies is the bridal chamber . . . the bridal chamber is within a realm superior to what we belong to and you cannot find anything like it . . ."; Robinson, *The Nag Hammadi Library*, 151.

16. *GBC* [4:2], 15–16; *MMU*, 93, 95.

17. *MMU*, 93-101.

made a journey to heaven in your light body; there you were born for the second time. This second birth was the first great stop to the light.

After Yeshua had cleansed the temple in Jerusalem of materialism and hypocrisy, he met the Pharisee Nicodemus, a leader of the people, who secretly visited him at night. In Miryam's text, Nicodemus said to Yeshua: "Rabbi, we know that you are a teacher come from God, for no one can do these signs that you do, unless the Spirit is with him."[18]

In the modern translations of John's Greek text, Jesus replied as follows: "Truly, truly I say to you, unless anyone is born anew, that person cannot see the kingdom of God." However, the original Greek text says: "Unless one is born from above (that is, born from the high womb of Spirit or the Mother) he cannot find the Kingdom."[19] The modern translations are not correct when translating: "Unless one is born anew." In Miryam's gospel, Yeshua said: "Unless one is born of the breath of the Spirit, then that one will not find the kingdom of God."[20] From all this it can be deduced that Yeshua and Miryam were reborn from above, from the womb and breath of Spirit.

The Second Step: The Anointing

In *The Gospel of the Beloved Companion* Miryam anointed the head of her husband Yeshua twice. It was a ritual act that the king underwent in Ugarit in Canaan. Later in Ancient Israel it was a ritual temple act in which the Mother's ointment opened the crown chakra further so more light could flow in.[21] In other words: the anointing is another step in the process of awakening on the way to the light.

We do not exactly know when Miryam—as the representative and earthly manifestation of the Mother—anointed Yeshua the first time,

---

18. *GBC* [9:1], 21–22; *MMU*, "The Year 31," paragraph 3, 143–49.
19. Jn. 3:3 and Jn.3:5
20. *GBC* [9:3], 22; *MMU*, "The Year 31," paragraph 3, "Nicodemus and the Second Birth," 143, 145.
21. Wyatt, "Royal Religion in Ancient Judah," 61–81 about the rituals during the crowning ceremonies.

but in the story of the resurrection of Lazarus (Miryam's brother), it is mentioned in passing that she as the Beloved Companion had previously anointed the head of Yeshua.[22]

After Yeshua resurrected Lazarus, he had to go into hiding because both he and Lazarus were wanted by authorities.[23] They fled to Ephraim. Yet, six days before the Jewish Passover, Yeshua returned to Bethany near Jerusalem, where he and his wife Miryam of Bethany lived. It was there that Miryam anointed Yeshua's head for the second time; she poured a jar of oil of pure and expensive spikenard over Yeshua's head,[24] causing the house to become full of the sweet smell of the oil. As said, after this he proclaimed her the Migdalah: the toweress or the exalted one.

Yeshua literally called her "the tower of the flock," which was again an old honorary title for Wisdom. In the language of the First Temple, Wisdom was compared to the City of Jerusalem, with ramparts and towers and especially to the temple tower.[25] The tower of the flock is a refuge, a safe place for the followers. In the ancient royal temple rite, the queen bride, in her role as high priestess, anointed her husband the king in his ascent to the light.[26]

The Third Step:
The Eucharist—Sharing in the Food of Wisdom

The word *Eucharist* literally means "thanksgiving." The early Christians enjoyed a meal of bread and water or wine together. These were symbols

---

22. *GBC* [28:1], 52; *MMU*, 299.
23. *GBC* [31:2], 59; Jn. 11:2; *MMU*, "The Year 33," paragraph 2, "Jesus Is Outlawed," 321–23.
24. *GBC* [32:2], 59; Jn. 12:3; *MMU*, "The Year 33," paragraph 5, "Mary Magdalene Anoints the Head of Jesus," 343–55.
25. In Rev. 21:9–12 the city of Jerusalem is female, she is a bride, the bride of the lamb; the city has high walls.
26. *GBC* [28], 52–55; cf. *MMU*, 343–55: about the anointment of the head of Yeshua and not of his feet; Beavis, *Rediscovering the Marys*, 232 with reference to Margaret Starbird and the Song of Songs 1:12; Baring, *The Myth of the Goddess*, n5.

of Wisdom to be found in the ancient Hebrew wisdom tradition. Lady Wisdom sent out her female temple servants with a dinner invitation to invite the people.

> *Come, eat of my bread*
> *and drink of the wine that I have mixed.*
> *Leave the simple ones, and live*
> *and walk in the way of insight.*[27]

In Miryam's gospel, Yeshua said: "My words are the bread of life . . ."[28] and "I am the bread which came from the Spirit."[29] In my opinion, giving thanks together and eating the food of Wisdom together has to do with the third step on the way to the light and enlightenment. Increasing light and increasing connection to higher worlds brings trust, gratitude, joy, surrender and abundance, symbolized here by bread and wine.

In Miryam's document the last meeting takes place on the evening before Yeshua's arrest; this is called Maundy Thursday in the Christian tradition. At the end of the meal, Yeshua announced that someone will hand him over, whereupon Peter beckoned to the Beloved Companion, who was reclining at the table and leaning against Yeshua's chest/heart, and asked her, "Tell us who it is of whom he speaks."

And she, still leaning back against Yeshua's chest/heart, asked him: "Rabbouni, who is it?"

Yeshua replied, "It is he to whom I will give this piece of bread when I have dipped it." It is unclear whether the liquid is water or wine. He dipped the piece of bread and gave it to Judah, son of Simon of Kerioth, and said to him: "What you must do, do quickly."[30] The Gospel of Miryam proves in this scene that Miryam, as the Beloved Companion, is present at this meal and is reclining with Yeshua.

---

27. Prov. 9:5; cf. Sir. 21:21; *MMU*, 205.
28. *GBC* [17:8], 34; *MMU*, 204.
29. *GBC* [18:1], 35; *MMU*, 209.
30. *GBC* [35:7], 63; Jn. 113:32; *MMU*, "The Year 33," paragraph 8, "The Last Meal," 374, 391–92.

At the end of a long speech, Yeshua gave his disciples a new commandment, in fact a new assignment. It meant: Love one another, just as I have loved you. By this everyone will know that you are my disciples, if you have love for one another.[31] After which Yeshua announced his departure.

The Fourth Step: The Redemption

The process of redemption begins with the crucifixion of all the selfishness and ego tendencies of the little lower self. These you lay down in order to rise in the light and get in touch with your higher self. The gnosis speaks of the "twin," and this refers to the connection that arises between the lower and higher self—no separateness. In fact, that is what is meant by the sacred marriage that takes place in the bridal chamber: it is an inner marriage between the two parts of the twins, the lower and the higher self. That inner marriage puts an end to inner divisions. The process of redemption actually involves letting the connection that makes the two into one grow. When you succeed in transforming your physical body to a large extent into a light body, you speak of resurrection. You seem to have transformed 75 percent of the subatomic particles into light. Now, only a quarter of the body consists of atoms; the rest is transformed into lightening subatomic particles.[32]

Yeshua was arrested in the night from Thursday to Friday and brought before Annas[33] and the high priest Caiaphas.[34] They handed him over to Pilate, who found no guilt in Yeshua and did not know how to deal with the situation. When he left judgment to the incendiary

31. *GBC* [35:9], 64; *MMU*, "The Year 33," paragraph 8, "The Last Meal," 375, 400.
32. Creme, *The Ageless Wisdom Teaching*; Dutch transl.: *De leringen van de Oude Wijsheid*, 1998, 2006 (2ᵉ ed.), 28.
33. *GBC* [36], 67–68; *MMU*, "The Year 33," paragraph 9, "The Arrest in the Garden of Olives," 403–07.
34. *GBC* [37:6], 68–70; *MMU*, "The Year 33," paragraph 10, "Jesus Questioned by Annas and Caiaphas," 409–13.

crowd, they called for Yeshua to be crucified.[35] Pilate handed Yeshua over to those people to carry out the sentence.[36] In her Gospel, Miryam was extremely brief about the crucifixion. Some researchers consider that Yeshua, by drinking a sponge with vinegar, is deliberately rendered unconscious. Immediately afterward Yeshua said, "It is finished" and bowed his head.[37] One of the soldiers heard this and pierced his side with a lance, and blood and water came out, a sign that he may still have been alive. Unlike the Gospel of John, this all happened very quickly in Miryam's Gospel. So it may very well be that Jesus was not yet dead. Before the coming Sabbath and before it was dark, the bodies had to be removed from the crosses. We know that after Jesus had bowed his head, it got dark very quickly. There is every reason to believe that Jesus was crucified only briefly and, though unconscious, was still alive afterward.[38]

Joseph of Arimathea went to Pilate and asked for permission to remove the body. Permission was granted, and Joseph took it away.[39] Nicodemus came with a large amount of healing herbs, a mixture of myrrh and aloes, possibly another proof that Jesus was still alive.[40] Contrary to the course of events in the Gospel of John, Miryam the Migdalah and Mary the mother of Yeshua took over tending to Yeshua's body by wrapping it in linen cloths with the herbs, "as the burial rite required."[41] Near the place where Yeshua was crucified there is a garden, and in that garden there was a new tomb, where Yeshua was laid down.

Miryam went to the tomb on the first day of the week, when it

---

35. *GBC* [38], 70–72; *MMU*, "The Year 33," paragraph 11, "Jesus Questioned by Pilate," 415–27.

36. *GBC* [39], 73; *MMU*, "The Year 33," paragraph 11, "The Crucifixion," 429–443.

37. *GBC* [39:5], 73; *MMU*, 430, 439.

38. Larson, *The Essene-Christian Faith*, 186–191 gives several arguments that Jesus is still alive after the crucifixion; Kersten and Gruber, *The Jesus Conspiracy*, 251; Schonfield, *The Passover Plot*, 2020.

39. *GBC* [40], 74–76; *MMU*, "The Year 33," paragraph 13, "Mary Magdalene, the First Witness to the Resurrection," 447–65.

40. *GBC* [40:2], 74; *MMU*, 447.

41. *GBC* [40:2], 74; *MMU*, 454, 457.

was still dark. She saw that the stone in front of the entrance had been removed. She stooped and looked inside and saw that the tomb was empty; the linen cloths were left where the body had been. She stood outside and cried. She heard a noise, turned around, and saw a figure standing nearby, but because she was crying, she didn't realize it was Yeshua until he said her name and then she recognized him. He said, "Miryam, do not hold to me, for I am not of the flesh yet neither am I one with the Spirit; but rather go to my disciples and tell them you have seen me."[42] Miryam's version is short and clear compared to the Gospel of John, who, because he wants to give the leading role of this episode to Peter, introduces all kinds of artifices into the text, making the text illogical. It is abundantly clear that John's editors later inserted text to emphasize Peter's role.

It is important to realize that the concepts of resurrection and ascension have become distorted and confused in mainstream Christianity. Emphasis has been placed on the uniqueness of the resurrection of the only begotten Son of God, who would have died for the sins of mankind.[43] That creates a great distance between this very special Son of God and the people. Yeshua wanted to show us the way, which goes through the crucifixion of the lower self to the resurrection in the higher self. Resurrection only gave access to continued existence in a much lighter physical body. The alternative sources stated that every physical body, even a resurrected body, must eventually die, while the Body of Light, the body that makes the ascension, is immortal.[44]

The Fifth Step:
The Ascension—The Church's Vision

The ascension of Jesus into heaven is celebrated in mainstream Christianity forty days after the resurrection (Easter) with Ascension

---

42. *GBC* [40:6], 75; *MMU*, 459.
43. Greene, "The Celestial Ascent," 12, 44, 49; see Jn. 3:16, *MMU*, 143n9.
44. Wilson and Prentis, *The Essenes*, 263–65.

Day. It is important to realize that traditional Christianity has placed a strong emphasis on the ascension of Jesus, which is said to be exclusive to him, despite the fact that Yeshua emphasized many times that he only wanted to be an example for others who follow him.

In the Roman Catholic Church, another very special person later makes an ascension: Mary, the mother of Yeshua. On icons you see how he holds her swaddled soul in his arms next to the bier containing Mary's dead body. Mary's soul made her ascension. This is celebrated in the Church on the Feast of the Assumption of Mary on August 15; the doctrine was declared a dogma in 1952. The Church had sidetracked the general phenomenon of the ascent into the heavens, instead emphasizing the unique ascension of Yeshua and Mary his mother.

But in the Hebrew First-Temple tradition it is different. Many prophets made a heavenly journey and received visions that they shared with those around them after returning to the physical body. In this tradition are the gnostics, who also made heavenly journeys, and *The Gospel of the Beloved Companion* belongs in this tradition. Miryam described her ascension extensively, and she undertook her ascent together with Yeshua, who guided her. Both went through the four stages. After the fourth phase of crucifixion and resurrection, the fifth and final phase—ascension—finally followed.

The Ascension of Yeshua: The Alternative Sources

As well as the apocryphal sources, the alternative sources have indicated that Yeshua survived the crucifixion. They don't say when Yeshua underwent the final ascension, but it is an *ascensio post-mortem*—an ascension after death. This was about the final transition to the Light World. That must have been somewhere in Asia at a much older age: he would have lived well into his eighties. The alternative sources agree with various historical sources from Central Asia, which are not recognized by conventional Christianity. Although the alternative sources did not provide information about Yeshua's final ascension, they provided information about the content of Yeshua's teachings about the way to the Light or the process of ascension. This teaching was also propagated by his designated successor, Mary Magdalene.

According to Daniel Benezra, Yeshua taught about the ascension in private within the Essene communities and in the inner circles of female and male followers. But he also publicly talked about ascension during trips to villages and towns. Yeshua called the ascended state "the Kingdom of Heaven," for it is a separate sphere with its own laws and its own state of being. The words *Eternal Life* are also used to indicate this state, for people will wear a Body of Light that is immortal and will not perish or die.[45] During the resurrection in the tomb, Yeshua still had an enlightened physical body, but after the final full ascension, according to Daniel Benezra, the physical form passed into the glory of the formless state.[46]

Daniel Benezra made it clear that the Way to the Light had different phases.[47] During the early stages of ascent and lightening, you can still live in a physical body. But when the task on Earth is completed, the physical body dissolves into Light. This is the basis of an ascension that takes place after death: the consciousness lives on forever in the Light Body or Merkabah. Daniel Benezra is convinced that Yeshua and Miryam had both definitively passed into the full ascended state at the end of their lives, but as mentioned, did not know when that was.[48] For both, it concerns an ascension post-mortem, the final ascension after death.

Alariel shared that ascension is a process that consisted of an initial and a final phase.[49] In the initial phase it was about raising the

---

45. Wilson and Prentis, *The Essenes*, 260.
46. Wilson and Prentis, *The Essenes*, 262.
47. Wilson and Prentis, *The Essenes*, 258, 291. The Melchizedek teachings give three stages: 1. Deep knowing or gnosis, then the boundaries between the individual and the other life forms disappear. 2. Integration; you learn to live this deep knowing in everyday life and integrate it, and you become a self- or God-realized being. 3. Ascension or the stage in which you merge from earthly reality into light.
48. Wilson and Prentis, *The Essenes*, 258.
49. Wilson and Prentis, *Beyond Limitations*, 115.

light frequency and slowly rising up into the Light. In this process, the physical body was increasingly absorbed by the Light, so the ascension was a process of lightening up and becoming lighter.[50] The vibration increased, the consciousness expanded, and the light in the physical body became stronger. This process culminated in the final phase of the ascension process: the ascension or the final transition to 5D, the world of nonduality. Until recently, this in most cases took place after death. When the task or mission on Earth was finished, you would die, and with sufficient luminosity, you would be passed to 5D. This concept is changing in this time of transition because of the increased vibrational frequency of the Earth. It is now possible to be on Earth in an ascended 5D form.

According to Daniel, there was disagreement about ascension even among the disciples during Yeshua's lifetime. Some thought that such a path was not practical, saying that human beings can only progress a little at a time, so there was no need to elaborate on such a distant goal. But a few, like John and James, Thomas and Mary Anna and Mary Magdalene agreed with Yeshua. They said that if you do not put the completion stage before the people, they will not be able to aspire to it and may spend many lives at lesser levels (and keep circling around in the loop of 3D and 4D). If the idealism, the Sacred Fire in the heart, is not kindled many people may sleep their lives away within the home, the market, or the temple.[51]

As Daniel reported, and I summarize, in the past, ascension was only achieved with great difficulty and after many lives of struggle, but Yeshua came to change that and to make the way clear and straight. He told us that ascension is about a transformation of the heart. A brilliant mind will not help you to ascend, but a loving heart will.[52] As the cycle progresses, it will become easier. When you reach the end of the cycle, it will be like stepping into the Light. It is important to see ascension

50. Wilson and Prentis, *The Essenes*, 340; Wilson and Prentis, *Power of the Magdalene*, 222.
51. Wilson and Prentis, *The Essenes*, 262.
52. Wilson and Prentis, *The Essenes*, 260.

as the completion stage of your existence. Yeshua said that we should aspire toward it in our lives; from day to day, moment to moment, we should focus upon our area of service, our self-appointed task. He also said that when our task upon Earth is completed, we should be ready to ascend into the realm of Light. Yeshua told us that there comes a point in the development of every human being when the soul is filled with a great longing for the Light.[53] Only the heart can open the door; it is unconditional love working through the heart which makes ascension possible.

This is a simpler and more direct way than was practiced before, and will enable many people to ascend who could not master the complex procedures I found described in the Egyptian texts.[54] To move forward in this way, we will need to give up every bit of hatred in our hearts. In this, I separated myself from many of the priests, who held that hatred towards the sons of darkness is a virtue, even a duty for the Essenes. I couldn't agree with this, for I saw the destructive effect that hatred had in people's lives.[55]

So, Daniel believed that it was ultimately about the knowledge of the heart. Book knowledge is only a means to this end.[56]

As the alternative sources have shown us, in the past, ascension was very difficult to achieve. Yeshua and Mary Magdalene came to change that. Alariel explained it this way. In Yeshua's time the vibration of the Earth was low; the distance between 3D and 5D was too great. A full ascension inevitably meant disappearing into Light and leaving the Earth's frequency since that frequency was so dense and so far from the vibration of the ascended state of being. Only a full ascension after death was possible.[57]

Alariel continued, and I summarize, that we have now come to the

53. Wilson and Prentis, *The Essenes*, 261.
54. Wilson and Prentis, *The Essenes*, 261.
55. Wilson and Prentis, *The Essenes*, 261.
56. Wilson and Prentis, *The Essenes*, 246.
57. Wilson and Prentis, *The Essenes*, 257–59; Wilson and Prentis, *Atlantis and the New Consciousness*, 125.

end of a planetary cycle, where all beings will be assessed according to the brightness of the Light within the heart.[58] Ascending becomes easier as the distance between the 3D reality and 5D reality becomes smaller. The veil is starting to thin as the planet nears transition.[59] Because of the Earth's current transition to the Light you can now ascend without losing your physical body. So, it is now an *Ascensio prae-mortem*. Ascension gives you a much richer earthly life while remaining in your physical body. It is a new option made possible by the development of human consciousness over the millennia. There are two options: after the full Ascension to 5D, you can choose to serve in other places in the Universe beyond Earth. But it is hoped that when most people reach Ascension, they will stay on Earth because they want to serve the Earth and help create a New Earth. So, you can truly choose to work in an ascended state for the New Earth.[60]

To recap: two thousand years of distortion of the original message of Yeshua and Miryam about ascension are now behind us. It grew dark on the Earth, yet a seed was planted two thousand years ago. Now that the seed has taken root, it can possibly grow under Miryam's direction into a Great Tree, the Tree of Life through which people will understand the true process of ascension.

58. Wilson and Prentis, *The Essenes*, 247, 257–59.

59. Wilson and Prentis, *The Essenes*, 258; Wilson and Prentis, *Atlantis and the New Consciousness*, 125.

60. Wilson and Prentis, *Beyond Limitations*, 115–16; Wilson and Prentis, *Atlantis and the New Consciousness*, 124, 141.

PORTAL FOUR

WHAT HAPPENED TO YESHUA AFTER THE CRUCIFIXION?

MARY·HATH·CHOSEN·THAT·GOOD·PART·WHICH
'SHALL·NOT·BE·TAKEN·AWAY·FROM·HER

Erected to the Glory of God in loving memory of
Mary Forrest of Ardow died 23rd October 1904
by her affectionate sister Isabella D. Forrest.
Watsonian of Ardow...

Figure 12.0. Jesus and Mary Magdalene holding hands according to an ancient Celtic custom, the handfasting ritual, a marriage ritual. Mary Magdalene is clearly pregnant. Stained glass panel from Kilmore Church (from Chil Moire or Church of Mary) in the Scottish town of Dervaig on the Isle of Mull, east of the Isle of Iona. The panel was created by the stained glass artist Stephen Adam from Glasgow and was installed around 1910. The small white church with the round tower of Mull is indebted to the abbey of Iona, the Iona Abbey, formerly called St. Mary's Cathedral; there would be a direct relationship between this church and the island of Iona.[1]

This image may be used without prior permission for any scholarly or educational purpose.
Credit: Stephen Adam, *The Victorian Web.*

---

1. van Dijk, *Maria Magdalena, The Lady of Glastonbury and Iona,* 177.

12
YESHUA SURVIVED THE CRUCIFIXION

This chapter examines the events after the crucifixion. It deals successively with the alternative sources, an Essene source, and sources from Central Asia. Both Muslims and Hindus indicate that Yeshua survived the crucifixion.

The Regression Accounts

During the crucifixion, the various communities of Essenes did everything they could to support Yeshua. According to Daniel Benezra, energy flowed through the network of crystals in the three Essene communities that made up the North Triangle, namely Carmel, Jenin, and Rama, and the three Essene communities that made up the South Triangle, namely Hebron, Ein Gedi, and Arad. In between was the Middle Triangle with support groups in Jerusalem, Qumran, and in the cave with the large crystal as an intermediary. In addition, there was also support from a crystal, which was carried by someone under the cross, without others seeing this.[2] Daniel Benezra resided in Hebron during the crucifixion; there he

2. Wilson and Prentis, *The Essenes*, 155–66; Wilson and Prentis, *The Magdalene Version*, 30–39 is about Sarah, daughter of Isaac, brother of Mary Anna, niece of Yeshua, and one of the female disciples of the first circle, who also studied in Carmel, and who stands with a crystal under the cross; Wilson and Prentis, *Power of the Magdalene*, 144–157, 153 about the support of a group from the Isis temple with crystals.

meditated with four others. The five of them were able to clairvoyantly perceive the crucifixion from a distance, again using crystals.

The process of keeping the energy flow to Yeshua was immensely difficult, because there were opposing forces at work that tried to block this. The energy process became heavier and heavier, and although it didn't last long by earthly standards, it seemed to take an eternity for those involved. When it became too heavy, some people couldn't keep it up. They sent out a cry for help to the core group meditating in the cave above Jerusalem and Qumran. They sent a request for help to the angels, to Archangel Michael. Then the rescue came, and then Daniel concluded that Yeshua had been able to complete the entire energy cycle. They followed extremely strict orders to immediately stop watching when Yeshua was taken down from the cross. In the support groups in the North and South Triangles, people had given everything they had. Daniel explained that Yeshua was not the only one who made a sacrifice. Two people in the support groups within the communities died, and some others remained ill for years afterwards, reduced to a shell of what they once were. It was such a battle.[3]

Regression records clarify that many Essenes were physically present at the crucifixion. Some Essenes watched from a hill; they did not allow themselves to become emotional at the spectacle; they had to keep up the love energy.[4] Joseph of Arimathea had his arm around his little sister Mary Anna. After that, things developed quickly and with great sadness in his heart, he had to leave his sister behind.

Earlier, I discussed the possibility that Yeshua did not die on the cross. The Essenes were masters of deliberate confusion around the mystery of the empty tomb. Suddi gave Dolores Cannon the explanation that Jesus's body went through an accelerated process of disintegrating to dust.[5] There were many other explanations circulating. Daniel

---

3. Wilson and Prentis, *The Essenes*, 164–67.
4. Wilson and Prentis, *The Essenes*, 167–68.
5. Cannon, *Jesus and the Essenes*, 266; Wilson and Prentis, *The Essenes*, 175; Larson, *The Essene-Christian Faith*, 187–89 lists several theories regarding whether Jesus survived or was dead and what happened to his body next.

Benezra said, "The news of the empty tomb was spreading fast. There was so much confusion at that time that frankly I did not know what to believe. But when I spoke to Joseph of Arimathea in the Interlife after his death, the truth finally emerged. He told me that the vanishing into dust was a rumor spread by the core group to protect Yeshua. To say that Yeshua turned to dust very quickly with the help of the angels was a most useful story. It explained why the tomb was empty and also discouraged anyone from looking for the body. The core group didn't want anyone looking for Yeshua, alive or dead, and his security had to be considered."[6]

What happened according to Daniel? Like Suddi, Daniel was unaware of the real events during his lifetime. The first thing he did when he arrived at the Interlife was to ask what happened to Yeshua. Daniel reported: "Those assigned to help me there laughed and said, 'All the Essenes are asking us this. Be patient, you will know soon enough.' And through the Angelic Record I was able to review it all. After Yeshua was taken down from the cross, it became clear that he was far out of his body, for he did not move at all. I saw him being placed in the tomb and the great stone being rolled across the entrance. I saw the light around Yeshua increase very strongly. One part of that light rose and ascended to the heavens, but the other part continued to focus around the body. Then I learned how well the core group had prepared for it all. His tomb was not really a tomb at all, but a healing chamber, into which healers could enter to do their work."[7]

When asked by Prentis how people could enter when a heavy stone blocked the entrance, Daniel replied:

The Essenes had constructed an tunnel leading from the back of the tomb to Joseph's house, which was nearby. That tunnel was cut through the rock over a limited distance and then sloped

---

6. Wilson and Prentis, *The Essenes*, 175–76, 204–205 about the many rumors surrounding Yeshua and the persecution that took place afterward.
7. Wilson and Prentis, *The Essenes*, 169–70.

down into the earth. Someone could walk in the tunnel, not in full length, but slightly bent. The tunnel led to a place behind the tomb. . .

The door into the tomb was ingeniously made of thick wood, but the outer side of it, which would be visible to anyone standing inside the tomb, was treated to resemble rock. The door hung in locks. When you went through the tunnel down to the tomb you had to lift the door out of the locks and that door could be taken out and put aside. Next, you crawled through the hole into the tomb. Anyone standing inside the tomb with the door hanging in the lock saw nothing but rock. Everything was very skillfully made. Understand that a lot of time was available to prepare the tomb. We knew that the Teacher would come and go through a ritual death experience. Therefore, the need for a tomb as a healing chamber was evident from the outset. Under the leadership of the Kaloo, the ancient ones, and the core group, our people had been working towards that moment. Of course, it was a covert operation. All work had to be done in great secrecy, and very few people within the Essene Brotherhood knew about it. In my lifetime, I never heard about it.[8]

Daniel continued that, after the light had hovered around Yeshua's body for some time, he saw him breathing more deeply, but even then it was clear that much needed to be done to revive him. Although he was far out of his body, the shock of the crucifixion must have been immense. Although Yeshua trained for years in body and mind to withstand this massive shock, its impact should not be underestimated. Luke (Yeshua's uncle and physician) and Clare (Yeshua's sister) were the most powerful healers in the entire Essene brotherhood, and they worked with him day and night. The large stone in front of the entrance to the tomb was designed in such a way that light from the outside was completely blocked. It had also been made impossible for people outside the

---

8. Wilson and Prentis, *The Essenes*, 170–71.

tomb to see the light from the lamps inside the tomb. And the big stone also muffled every sound inside.[9]

Daniel went on saying that he was shown what happened during the healing process. He saw the healers anointing Yeshua with healing ointments and there were huge crystals around him. This was how he was brought back to life. It took some time, because he was far out of his body. But the healers were convinced that they could bring him back. And when he came back into his body, it happened slowly, like someone who had drifted away to a place very far away. He came reluctantly, because there was pain from the wounds. Even with the help of all the healing therapies, it was an intense experience. He gasped in pain as he returned to his body. He opened his eyes and saw Clare (his sister) smiling as she looked down at him, and tears of joy ran down her face.[10]

During another session, Prentis asked Joseph of Arimathea how Yeshua came out of the tomb. Joseph responded: "After we had resuscitated Yeshua, he was not able to walk, so we used a simple stretcher and slid him into the tunnel, and took him through it. He was lifted out at my house, which was close by, but still it was a difficult and frightening operation. I felt the weight of the responsibility, of all the secrecy and preparation. Only a few members of the Essene brotherhood knew the scope of our plans, and the tunnel's existence was known only to those directly involved. Not even Mary Anna (Jospeh's sister) was aware of the tunnel while it was being made. I didn't tell her until immediately before meeting in the tomb, for the fewer who knew the safer it was."[11]

Prentis asked Daniel about this, and he replied that:

Joseph knew that the Pharisees would examine the tomb after the body of Yeshua had disappeared. That was why Joseph—in all the confusion—had to act quickly. A prominent Essene in Qumran had

---

9. Wilson and Prentis, *The Essenes*, 172.
10. Wilson and Prentis, *The Essenes*, 172.
11. Wilson and Prentis, *The Essenes*, 172–73.

recently died, and he was hastily laid to rest in the tomb. The stone was rolled back, and the work really began. A stable was hurriedly built on Joseph's land behind the tomb, right above where they knew the tunnel ran. While the stable was used as a cover, they tore down the tunnel and filled it with earth. In doing so, they removed every trace of the carefully crafted door. By the time the Pharisees demanded to see the inside of the tomb, they had to ask permission from the Sanhedrin because the Essenes, led by Joseph, spoke in great anger of desecrating the tomb.

. . . By the time the Pharisees finally got permission to open the tomb and inspect it, the work was done and nothing suspicious was found. As usual, the brotherhood had covered the tracks very efficiently, and by that time Yeshua had long since been taken to Qumran and thence from one safe house to another north of Damascus. After he had safely left our country, the brothers began to transfer the most precious possessions from Palestine via the northern route to places outside Palestine. First, they moved the most valuable scrolls. We knew it was only a matter of time before Qumran would be destroyed and our main task would be done by then. When the outer work of the Teacher of Righteousness was finished, the whole structure around him could be dismantled.[12]

The Essene Letter:
Yeshua Survived the Crucifixion

There is an Essene letter from around 40 CE that stated Yeshua survived the crucifixion. The original had been lost, and the modern translations were based on a Latin translation, which makes it uncertain whether it was an authentic text. In the letter an Essene brother spoke; he claimed to have been an eyewitness to the crucifixion. In his account, Joseph of Arimathea played a major role in the events surrounding the crucifix-

---

12. Wilson and Prentis, *The Essenes*, 173, 174.

ion. The Essene brother wrote that Joseph was wealthy and a member of the council and that—although he appeared to belong to no party—he was "secretly a member of our holy order and lived by our laws. His friend Nicodemus was a very learned man and ranked among the highest ranks of our order. He knew the secrets of the therapeuts [Essene healers in Egypt] and was often with us."[13]

I will briefly summarize the letter: There was deep darkness, and a hissing sound, the kind that always precedes an earthquake, was heard. When the earthquake followed, the mountain started to shake and the surroundings, with city and temple, began to shift back and forth. Many of those present fled in panic.[14] When Joseph and Nicodemus arrive at the cross, it seems strange to Nicodemus that Jesus, who had "been on the cross for less than seven hours," should already be dead.[15] With Yeshua seemingly dead, his Roman exectioners did not break his bones, as this was routinely done during crucifixion to expedite death.

The Essene eyewitness reported: "One of the soldiers thrusts his spear into the body in such a way that it slides over the hip into the side, and this is taken by the centurion as a sure sign that he is actually dead; then he hurries off to make his report. But from the insignificant wound flows blood and water; John marvels at this, and my hopes are revived, for even John knows from the knowledge of our brotherhood that nothing but a few drops of thickened blood flow from a wound in a dead body, but now both water and blood flow from it."[16]

Joseph went to Pilate to obtain the body. Pilate was favorable toward him and allowed him to take over the body without having to pay for it; this was unusual and would later lead the Pharisees to think that Joseph and Pilate were in cahoots.[17] In the meantime Nicodemus had obtained a lot of herbs.

Together they took him down from the cross and laid him down

13. Anonymous, *The Crucifixion—By an Eye-Witness*, 66.
14. Anonymous, *The Crucifixion—By an Eye-Witness*, 65.
15. Anonymous, *The Crucifixion—By an Eye-Witness*, 67.
16. Anonymous, *The Crucifixion—By an Eye-Witness*, 71.
17. Anonymous, *The Crucifixion—By an Eye-Witness*, 75–76.

on the ground with great care. The Essene brother wrote: "Nicodemus spreads out long strips of byssos [a vegetable fiber used to make linen] and puts the strong herbs and healing ointments he has brought with him on them. Their effect is known only within our order. He binds this around Jesus's body, pretending to keep the body from falling apart because he could only embalm it after the festival (meaning the Passover). These herbs and ointments have great healing powers."[18] The body was placed in the rock tomb that belonged to Joseph. They burned aloes and other invigorating herbs in the cave, and while the body rested on a bed of moss, still stiff and lifeless, they placed a large stone before the entrance so that the odors could better permeate the cave.[19] Thirty hours later and after another earthquake, they went to the tomb and discover that Jesus was alive.[20] They offered him some dates and wine. After this, Jesus was refreshed and stood up on his own. Due to the ointments, the wounds on the hands already began healing. After the byssos cloths were removed, Joseph says that this is not the place to stay any longer. "But Jesus is not strong enough to walk far. Therefore, he is taken to the house belonging to our order near Calvary, in the garden, which also belongs to our brothers."[21]

The Letter of the Essenes continued: "After this, Jesus goes to his disciples and appears to them in his physical body. He begins to travel again and gather with his disciples. The pressure on Joseph of Arimathea is mounting.[22] Caiaphas spreads the rumor that the disciples stole the body of Jesus.[23] Joseph of Arimathea is arrested, but later released through the mediation of the Essenes.[24] Jesus meets again with many of his disciples but withdraws because it becomes too dangerous.

18. Anonymous, *The Crucifixion—By an Eye-Witness*, 73–74.
19. Anonymous, *The Crucifixion—By an Eye-Witness*, 75.
20. Anonymous, *The Crucifixion—By an Eye-Witness*, 79.
21. Anonymous, *The Crucifixion—By an Eye-Witness*, 81.
22. Anonymous, *The Crucifixion—By an Eye-Witness*, 100.
23. Anonymous, *The Crucifixion—By an Eye-Witness*, 100.
24. Anonymous, *The Crucifixion—By an Eye-Witness*, 110, 113. See also chapter 4, the apocryphal sources.

He is going to die from exhaustion six months after the crucifixion. He is buried at the Dead Sea according to the law of our brotherhood."[25]

The Essene's letter is popular, including among Islamic researchers who argue, partly on the basis of the Qur'an, that Jesus did not die on the cross.

Gnostic Sources Deny the Crucifixion

Certain gnostic writings oppose the Orthodox creed, which reads: "I believe in Jesus Christ, who suffered under Pontius Pilate and died and was buried. . . . In their view Christ brings salvation through higher knowledge and understanding. They see a difference between the invulnerable Christ, the divine savior, and the suffering Jesus. Docetism is popular in various variants among gnostic schools; some see the suffering of Jesus as a "puppetry," for a God-sent Christ, the bringer of revelation, who could not suffer.[26] The denial of crucifixion plays a central role in various gnostic writings.[27] I give an overview in the accompanying note.[28]

After the Crucifixion Yeshua Went Abroad:
The Alternative Sources

Prentis asked Alariel, "What did Yeshua do after the crucifixion?"

Alariel replied, "It was necessary for him to leave Israel. He could not stay because many people wanted to harm him, and he could no longer teach openly. He secretly went to Damascus and from there he

---

25. Anonymous, *The Crucifixion—By an Eye-Witness*, 127–28.
26. The redemption of the fallen parts of light does not take place through Jesus's suffering but through the transfer of higher knowledge or gnosis that leads to liberation.
27. Luttikhuizen, *De veelvormigheid van het vroegste christendom*, Intro., 11 and 143–150.
28. The Apocalypse of Peter (VII.3), The Letter of Peter to Philip (NHC VIII.2 and codex Tchacos); The Tripartite Tractate (NHC I.5); The First Apocalypse of James (NHV V.3 and codex Tchacos); The Second Treatise of the Great Seth (NHC VII.2); the Gospel of Judas (codex Tchacos); The Acts of John (chap. 99–101).

followed the ancient trade route eastward. That route took him through cities that would be called today Baghdad, Tehran, and Kabul, and after a long journey he finally reached India."[29]

Prentis asked Alariel again, "Did Yeshua spend some time on the island of Cyprus on the way to Damascus?"

Alariel said that, yes, Joseph of Arimathea had a large estate there, and it was the ideal place for Yeshua to rest and heal before embarking on the long journey eastwards. Joseph had made Cyprus the center of his operations in this area, a wise decision given the uncertain situation in Israel. From Cyprus he could control the shipping of tin to all Mediterranean ports where the Romans were present.[30]

Alariel continued:

Yeshua had already studied in India prior to the public years, partly with a being you know as Babaji.[31] On his return to India, he renewed ties with Babaji. He told Yeshua that the guru-apprentice system in India was degenerating. Because of greed and selfishness (of the gurus), many good aspirants did not get the keys to empowerment; thus, they were kept in a subservient role instead of rising in turn to independent guru status. Yeshua took it upon himself to revitalize the old system. He did this by traveling around India and visiting the ashrams. In doing so, he taught the gurus and restored the system to its original purity. He did this work partly in collaboration with the disciple Thomas, who had reached India by a different route (by sea) and who joined him there. And besides purifying the ashrams, they purified the energies of a number of holy places. This was important because those places were an integral part of the Hindu tradition.[32]

---

29. Wilson and Prentis, *Power of the Magdalene*, 106.
30. Wilson and Prentis, *Power of the Magdalene*, 106.
31. This is about Haidakhan Babaji, as mentioned by Paramahansa Yogananda, *Autobiography of a Yogi*, Los Angeles: Self-Realization Fellowship, 1974, chap. 33.
32. Wilson and Prentis, *Power of the Magdalene*, 106.

That the apostle Thomas worked in India is described in the apocry-phal Acts of Thomas. Here he worked together with Yeshua. Although it is tempting to follow the trail of Joseph of Arimathea, we'll do that later. Here I choose to follow the trail of Yeshua to the East and ask the question: To what extent do Eastern sources confirm the regression material?

AN ASIAN INTERLUDE: YESHUA'S JOURNEY IN ASIA

That Yeshua would not have died on the cross can be found in the long Qur'an sura 4 entitled "About the Women," in verse 157:

> And for their saying, "Indeed, we have killed the Messiah, Jesus, the son of Mary, the messenger of Allah." And they did not kill him, nor did they cause his death on the cross, but it was made to appear to them as such. And certainly, those who differ over it are in doubt about it. They have no knowledge of it but follow a conjecture and they killed him not, for certain.[33]

This passage probably refers to the Jews of Medina who claimed that they (the Jews) had crucified Yeshua. The imme-diately following verse 158 indicates that the Qur'an sees Jesus as an important prophet: "Nay, Allah exalted/raised him to His Presence. And Allah is ever Mighty and Wise."[34]

The tradition that Yeshua traveled in Central Asia after the crucifixion is strongly emphasized in the circles of Ahmadiyya Muslims. They cite various facets of the passion story that sup-posedly show that Yeshua survived the crucifixion. They base this on the canonical and gnostic gospels, the Letter of the

33. The Qur'an, sura 4, verse 157; Ahmad, *Jesus in Heaven on Earth*, 228.
34. The Qur'an, sura 4, verse 158; Ahmad, *Jesus in Heaven on Earth*, 228.

Essene brother, and the Qur'an. The following aspects are seen as proofs of Yeshua's survival.

1. Yeshua fell into a coma from an administered drink that stunned and rendered him unconscious.
2. The wound in his side was not serious; water and blood flows out, a token that he was still alive.
3. Nicodemus took with him large amounts of healing herbs; the wounds of the Essene Yeshua were cleansed and treated with healing herbs and ointments by healers from the Order of the Essenes.
4. The revitalized Yeshua was taken away, and this (and nothing else) explains why the tomb was later found empty.

In 1899, Hazrat Mirza Ghulam Ahmad, founder of the Ahmadiyya movement within Islam, proclaimed that Yeshua did not die on the cross and wrote a book about it titled *Jesus in India*. Yeshua is said to have escaped death and secretly left Palestine. He went in search of the lost tribes of Israel: ten tribes from the Northern Kingdom are said to have been carried away to the East after the Assyrian conquest of this area. The author referred to both the Bible and the Qur'an; this theme recurs in almost all of his eighty books, written in Persian, Urdu, and Arabic. It was a historical fact that King Sargon II occupied the Northern Kingdom or kingdom of Israel in 721 BCE and took many people into exile; they eventually settled in Iran, Afghanistan, and India.

Many of the founder Ahmad's followers continued to develop this theme after his death. From the early 1940s, lawyer Khwaja Nazir Ahmad started collecting detailed evidence. He found evidence of the relationship between the Jews and the people of Kashmir in historical, cultural, ethnological, linguistic, and religious terms. He wrote articles that culminated in a book, first published in 1952 and since then reprinted several times. A few years ago I received this book, *Jesus in Heaven on Earth,* as a birthday present from my son.

When the Muslims' view of the Prophet Isa/Issa or Yeshua comes up, I take up this voluminous and well-crafted book. It runs 470 pages and remains the most comprehensive treatment of the subject of Yeshua in Asia. The book serves as a treasure trove for later modern books about this subject. Parts of the book have been translated into German, French, Italian, and Arabic. The French Academy awarded the author the degree of doctor of literature. Once I start reading in this book, I am fascinated every time by the wealth of texts in the many Eastern languages. I am always impressed by the archaeological sources and the imprints of Yeshua's and Mary's names in landscape and city names. The information is so overwhelmingly beautiful, abundant, and rich that you can only be grateful that the author took the trouble to arrange it all in an orderly and readable way. At the end of the book, a long survey is given of people in East and West who, from 1890 to the present day, have been concerned with the question of whether Yeshua could have survived the crucifixion and made his way to Asia.[35]

Khwaja Nazir Ahmad put forward with many arguments, including medical ones, that Yeshua survived death on the cross and moved eastward. He referred to the Essene letter discussed above. As discussed earlier, Yeshua wanted to continue the mission among the "lost sheep of Israel," which were the ten tribes from the Northern Kingdom, who after its fall in 711 BCE had been transported to very far Eastern countries. The author argued that the people of Afghanistan and Kashmir were among the ten lost tribes of the Israelites;

35. Ahmad, *Jesus in Heaven on Earth*, appendix 1 by Nasir Ahmad, "A Brief Chronological Survey of Research," 451–459, 451.

Yeshua traveled to this area to continue his mission among the Israelites and died in Kashmir. His tomb was located in Srinagar, the capital of Kashmir, where it can be found to this day (see fig. 12.1).

Originally twelve tribes of Israel resided under the royal house of David. But after Solomon, his empire split under his son in 975 BCE into the Northern Kingdom of Israel, where ten tribes lived, and the southern kingdom of Judah, where two tribes lived. Since the fall of the Northern Kingdom in 711 BCE, ten of those twelve tribes of the Israelites had been carried away, and only the tribes of Judah and Benjamin remained in Judea in the south.[36] The word Jew originally means "descendant of Judah," the son of Jacob and belonging to his tribe and territory.[37] The ten lost tribes spread through Mesopotamia, Persia, Afghanistan, and western China.[38] Margaret Barker mentions the work of the Reverend Thomas Torrance, who in 1916 "discovered" that the Chiang Min, a Chinese population group living on the border of Tibet and China, "preserved the Israelite way of life and belief." In 1937

Figure 12.1. Map from Khwaja Nazir Ahmad, *Jesus in Heaven on Earth*, 382.

---

36. Ahmad, *Jesus in Heaven on Earth*, 286–297.

37. Ahmad, *Jesus in Heaven on Earth*, 286–297, 287.

38. 2 Kings 17:6; 18:11 where the text says that the king of Assyria took the Israelites and settled them in Assyria, Mesopotamia and Media; see Ahmad, *Jesus in Heaven on Earth*, 288n6 and 291n1 for reference to 2 Esd. 13:36–39.

he wrote a book about it. His son, the professor T. Torrance (who accompanied his father as a child and visited this border area), republished his father's book and drew attention to it again in his preface.[39]

The Israelite and Hebrew background of peoples in Central Asia was noticed by more people.[40] The German missionary Joseph Wolff, searched for the lost tribes of Israel from 1827–1834. He made extensive travels in Asia about which he wrote reports, and as a result, he was nicknamed "the missionary of the world." In about 1850, he described his journey to Bukhara in Uzbekistan, where a tribe resided who were descended from Joseph and called themselves the Bani Israel or Children of Israel; their language resembled Hebrew.[41] He was shown a scripture in Hebrew. These people did not call themselves Jews, a name associated with the kingdom of Judah, but Israelites, the common name for members of the twelve tribes. That Yeshua went in search of the lost sheep of Israel is not a myth but a historical fact; it appeared in Jewish texts[42] and in the canonical[43] and apocryphal texts of the New Testament.

According to Ahmad, Yeshua eventually traveled to Kashmir in physical form. To this end, he crossed several countries. The

39. Barker, *The Mother of the Lord*, 23–25.
40. Ahmad, *Jesus in Heaven on Earth*, 296 gives several references and remarks that the Jews west of Persia called themselves *Yahoodi* ("Jews") while the Jews east of Persia called themselves *Bani Israël* ("children of Israel").
41. Joseph Wolff, *Researches and Missionary Labours Among the Jews and Mohammedans and other Sects*, London: Nisbet & Co, 1835; Ahmad, *Jesus in Heaven on Earth*, 296n3.
42. The fate of the ten tribes is discussed by the sages of the Mishnah and Talmud. Their opinions vary: Rabbi Akiba believes the ten tribes will not return; Rabbi Eliezer, however, maintains that the ten tribes will eventually return (Mishna, Sanhedrin 10:3; for additional references see Babylonian Talmud, Shabbat 147b, and Numbers Rabbah 9:7). The supposed location of the ten lost tribes is in itself a subject of much speculation and legend. References to this theme are found in Jewish classical texts (Genesis Rabbah 73:6; Sanhedrin 10:6/29b). The legend is also mentioned by Flavius Josephus in *Jewish War*: 7:96–97 and by Pliny the Elder in *Historia Naturalis* 31:24.
43. Jn. 10:16 and Mt. 13:57

journey led first to Damascus in Syria, then via Aleppo he followed the old caravan route to Edessa in Eastern Turkey. That journey took four days before it continued East. In particular, Khwaja Nazir Ahmad's book *Jesus in Heaven on Earth* served as the basis for numerous Western authors who became fascinated with the subject. I mention Andreas Faber-Kaiser; Holger Kersten; Elizabeth Clare Prophet; Fida Hassnain; Michael Baigent, Richard Leigh, and Henry Lincoln; Hugh Schonfield; Ian Wilson; Suzanne Olsson; and a host of others.

The Journey of Yeshua to the East in Seven Stages

According to Khwaja Ahmad the following stages can be distinguished in Yeshua's journey to the East.

1. From Palestine to Damascus in Syria

After his crucifixion, Yeshua first traveled in physical form to a place more than three kilometers outside Damascus. This place is still named after him and reads *Maqam-i-Isa*; the place was originally called Rabwah. There he met Ananias and others who become his disciples.[44] He also met Saul, who was converted by him.[45] He received a letter from the king of Nisibis (which later became an important center of Syrian Christianity and is located east of Edessa in present-day Turkey). The king was sick and asked Yeshua to come to Nisibis to heal him. Jesus answered in a return letter that he was sending someone and would come later himself. A commission arrived from Jerusalem to arrest the converted Paul.[46] Now it was also becoming too dangerous for Yeshua in Damascus, so he decided to go to Nisibis himself.

---

44. Acts 9:10; Ahmad, *Jesus in Heaven on Earth*, 385.
45. Acts 9:4; Ahmad, *Jesus in Heaven on Earth*, 385.
46. Acts, 9:23–25; Ahmad, *Jesus in Heaven on Earth*, 385.

2. From Damascus to Nisibis

Nisibis or modern Nusaybin is located near ancient Edessa or modern Şanlıurfa on Turkey's eastern border. Yeshua knew that there were some lost tribes living in Nisibis, as mentioned by the author Flavius Josephus.[47] Ahmad gives three Islamic sources in which the journey of Yeshua to Nisibis was mentioned. In the second source it was mentioned that his mother Mary was with him, that he wore white clothes and a turban on this journey, and that he traveled with a staff in his hand while walking. This was repeated in the third source.[48]

3. From Nisibis to Iran

Yeshua left Nisibis and yet a fourth source gave the reason why: the people of Nisibis wanted to kill him;[49] therefore he decided to leave. From that moment on Yeshua traveled incognito under the name Yuz Asaf. *Yuz* stands for *Yusus* or Jesus, and *Asaf* is Hebrew for "collector of the lepers whom he healed." That his name consisted of amalgamations of Hebrew words was mentioned by four Islamic sources.[50] After that, there are sources that indicate that Yuz Asaf reached Persia or present-day Iran through the west. He preached here, and many believed in him.[51] A saying of Yuz Asaf was carved into a city wall, possibly the city of Kashan. It read: "The palaces of kings are devoid of three virtues: Wisdom, Patience, and Religious Wealth."[52]

---

47. Josephus, *Jewish War*, 18, 9:1–8.

48. Rashīd al-Dīn Tabīb and W. M. Thackston, *Jami-ut-tawarikh: A History of the Mongols*, vol 2:81, Boston: Harvard University Press, 1998–1999; see Ahmad, *Jesus in Heaven on Earth*, 385n5; this is repeated in the *Nasikh-ut-Tawarikh* vol. 1:149, see *Jesus in Heaven on Earth* 385n6.

49. Tafsir Ibn-i-Jarir at-Tabri, vol. 3, 197; *Jesus in Heaven on Earth*, 385n8.

50. Ahmad, *Jesus in Heaven on Earth*, 386n4–7.

51. Agha Mustafai, *Ahwali-i-Paras*, 219, see Ahmad, *Jesus in Heaven on Earth*, 387n1.

52. 'Arustu.' On Yuz Asaf in *Ma'arif*, Vol. 34:1, 1934, p. 17, see Ahmad, *Jesus in Heaven on Earth*, 387n2.

4. From Iran to Afghanistan

After that Yeshua can be traced in Afghanistan. Near Kabul he is said to have stopped at a pond, washed his hands and feet, and rested for a while. That pond still exists and is called The Pond of Issa in the Arabic book, *Tariq-a-Ajhan*.[53] In Ghazni in western Afghanistan and in Jalalabad in the far southeast, there are two holy places or *Ziarats* that bear the name of Yuz Asaf, for he was there and preached there.[54]

5. From Afghanistan to India

According to the Acts of Thomas, Yeshua and Thomas met sometime in 48–49 CE in Taxila, located in ancient India. The old name of Taxila is Takshashila. This city has hosted a university that is considered the oldest in the world; today it is an important tourist attraction. Since 1980, the place has been on the UNESCO world heritage list. The city is located on the northern headwaters of the Indus River on the eastern riverbank, then in ancient India, now in modern Pakistan.

Thomas, unlike Yeshua, did not travel by land but by sea to India. He traveled by boat from a port in Mesopotamia and went ashore in the city of Andropolis, at the mouth of the Indus. He traveled upstream to a place called Attock and was introduced to King Gondaphares.[55] This was narrated in the apocryphal Acts of Thomas, a text that also included the three important hymns about the Holy Spirit as Mother, as well as the famous Song of the Pearl. In Taxila in India, King Gondaphares reigned between 21 and 50 CE (see fig. 12.1, map) Thomas and Yeshua attended a royal wedding of the king's daughter here.[56] In chapters 11 and 12

53. Prophet, *The Lost Years of Jesus*, 45n71 with ref. to Abhedananda, Kashmir O Tibete, 236.
54. Ahmad, *Jesus in Heaven on Earth*, 387.
55. Ahmad, *Jesus in Heaven on Earth*, 368, 387.
56. Acts of Thomas chap. 9 and 10; Klijn, *Apocriefen van het Nieuwe Testament* II, 104–06; Ahmad, *Jesus in Heaven on Earth*, 368, 387, 419.

of the Acts of Thomas, Yeshua appears to the bridal couple, who had only previously seen the apostle Thomas, and told them, "I am not Judas, who is also called Thomas, but I am his brother." According to Ahmad, this passage tells us that Yeshua was the (twin) brother of Thomas (the name Thomas means "twin"), so Jesus, the twin brother of Thomas, was said to be physically present here.[57]

6. From Taxila, India, to Muree, Pakistan

When an invasion threatens the area around 50 CE, the group from Palestine seek refuge in the mountains close to Taxila on the Indus. One traveled through Mari, present-day Murree, which is forty-five miles away from Taxila (see fig. 12.1, map). Until 1875, this city was called Mari, after Mary, who traveled with the group. The name Mari is used for Mary by Afghans, Jews, and people from Kashmir.[58] Mari or Murree was located high in the mountains, which offered coolness in the summer and lots of snow in the winter. From Murree, located in present-day Pakistan, you have a view of the white snow-capped mountains in Kashmir. In Mari-Murree, Mary died.[59] Her tomb, of which Khwaja Ahmad provided many photos, was located next to the defense tower, which was built in 1898. Her tomb is still called *Mai Mari da Ashtan* or "the resting place of Mother Mary" to this day.[60] The tomb was restored in 1950 after the involvement of the author.

7. In Kashmir

Ahmad wrote: "We can almost say with certainty that Yeshua entered Kashmir through the valley called the *Yusu Margh*; this

---

57. Acts of Thomas chap. 11; Klijn, p. 106.
58. Ahmad, *Jesus in Heaven on Earth*, 369n1 and Wikipedia see "Murree."
59. Ahmad, *Jesus in Heaven on Earth*, 369, 387.
60. Ahmad, *Jesus in Heaven on Earth*, 369, images on 372, 387n4.

valley was actually named after him, and you can still find the tribe *Yadu* (Jews) there."[61] The author gave a list of topographical names where the names *Issa* and *Yusu* appeared separately; *Issa* appears in ten names and *Yusu* in eleven. Ahmad argued that not only the geography but also the history of Kashmir proves that Yeshua was there, probably between 60 and 87 CE.[62] There are said to be twenty-one historical records of Yeshua's stay in Kashmir. At the end of all this information, the author concluded: "The best proof of the presence of Yeshua in Kashmir is the existence of his tomb in the Mohalla Khaniyar in Srinagar, the capital of Kashmir."[63] The author devoted a separate chapter to this and provided numerous photographs of the outside and inside of the shrine.[64]

The Tomb of Yeshua in Srinagar

According to old charters, the funerary monument was erected around 112 CE. The traveling staff of Yuz Asaf is also kept there. Yuz Asaf was buried by Thomas according to Jewish custom in an east–west direction. He would have died a natural death. He may have lived to be eighty-one and died in 77 CE. According to Ahmad, Yeshua accomplished all this in his physical body. Only after physical death did he definitively make his full ascension.[65]

As mentioned, there are more than twenty-one historical documents attesting to the existence of Yeshua in Kashmir, where he was known both as Yuz Asaf and as Issa.

Among those twenty-one sources is the *Bhavishya Purana*, written in Sanskrit and possibly dated very early, around

61. Ahmad, *Jesus in Heaven on Earth*, 387n5.
62. Ahmad, *Jesus in Heaven on Earth*, 388, 415.
63. Ahmad, *Jesus in Heaven on Earth*, 388.
64. Ahmad, *Jesus in Heaven on Earth*, chap. 25, "The Tomb of Jesus," 396–416, with several photos of the staff.
65. Prophet, *Mary Magdalene and the Divine Feminine*, 280.

115 CE. This text contains an account of *Issa Masih* or Yeshua the Messiah.[66] It described the arrival of Yeshua in the Kashmir region of India and his encounter with King Shalivahana, who reigned in the first century and who received Yeshua as a guest for some time. A conversation ensued between Yuz Asaf and the raja (king). In it, Yuz Asaf said he had come to purify the religion. When the raja asks him what his religion was, the closing sentences of Jesus's impressive reply were: "It is to bring love, truth and purity into the heart, and that is why I am called Issa Masih" (Yeshua the Messiah).[67]

The tradition that Yeshua survived the crucifixion lives not only among Muslims, but also among certain groups of Hindus in India.[68] Although the Gospel of Issa stated that Yeshua did not survive the crucifixion, stories circulated in India that he did and went back to travel in India after the crucifixion. There are many written sources, apart from the oral tradition. In 1934, Jawaharlal Nehru wrote in his *Glimpses of World History*: "Through central Asia, in Kashmir, Ladak, and Tibet, and further north, there is a strong belief that Jesus traveled there."[69]

66. Ahmad, *Jesus in Heaven on Earth*, 402, 409n1, 419, 424.
67. Hassnain, *Search For the Historical Jesus*, 191.
68. The Natha yogis of northwestern India have a sacred scripture, *The Nathanamavali*, with passages about Ischa or Issa showing that he did not die on the cross but went into samadhi and afterward went to the land of the Aryans (India).
69. Klink, *De Onbekende Jezus*, 85.

What Happened to Joseph of Arimathea after the Crucifixion?

Figure 13.0. Tin merchant Joseph of Arimathea arrives on the island of Avalon. Saint Patrick's Chapel in the territory of Glastonbury Abbey. My photo, summer 2017.

13
JOSEPH OF ARIMATHEA IN ENGLAND

In this chapter, the alternative sources are discussed first, followed by the historical sources. Do they confirm the alternative sources?

The First Weeks after the Crucifixion: The Regression Accounts

Prentis asked Daniel where Joseph of Arimathea was during the first weeks after the crucifixion, which were full of confusion and chaos. Daniel replied:

> Joseph's intervention to obtain the body of Yeshua from Pilate, and to keep the tomb closed later, cost him dearly. This was the first time that he had publicly spoken out in favor of the Essene cause, because until then everything had been handled quietly behind the scenes. People were now beginning to wonder where Joseph's real loyalties lay. Finding nothing remarkable inside the tomb, the Pharisees turned their full wrath on Joseph. He realized that he had to leave. From then on all the investigations of the Jewish religious authorities were centered on Joseph and his connection with the Essenes. . . .
>
> The Pharisees were beginning to suspect that Joseph was more deeply involved with the Essene communities and the group around

Yeshua than they had thought. Because of his wealth and power, Joseph had not been questioned before. Now they were going to look into it more closely and all those journeys he had been making. The moment came when the Sanhedrin wanted to question him, but Joseph refused to respond. He hastily left Palestine on one of his ships. The Pharisees wanted an answer to their questions about the locations of the Essene community and who the real leaders were. By leaving the country, Joseph protected the brothers, the core group and the elders. Yet some—also among the Essenes—spoke ill of Joseph despite the fact that he had helped so many people. Many young students enjoyed free travel on his ships and continued their studies in England, for example.[1]

Alariel provided the following information: "When Joseph wanted to leave in great haste, he first went to his great friend among the Roman administrators to arrange an escort of a centurion and six Roman soldiers. Within the hour Joseph left and he went alone. Very early the next morning, Joseph's family and his staff left; they left with the Roman escort. They took all Joseph's files with them in their luggage. These were documents that would have greatly interested those in charge of the Sanhedrin investigation of Joseph and his connection to the group around Yeshua and would have roused ever greater suspicion. As part of the investigation, a representative of the Sanhedrin announced his arrival at Joseph's house in the early afternoon. To his great surprise, he found Joseph's house abandoned. The only one present was Joseph's friend Nicodemus. He showed the Pharisee around the deserted house and empty office with great courtesy and a knowing smile."[2]

When Prentis asked Joseph of Arimathea where he had gone, he replied as follows: "I had to leave very quickly. There was no time to take Mary (his younger sister) with me, for the high priest's servants were hot on my heels. I hurried to the coast, boarded one of my ships

---

1. Wilson and Prentis, *The Essenes*, 174, 204–5.
2. Wilson and Prentis, *The Magdalene Version*, 47–48.

and left in great haste."[3] In this regression session, Joseph showed his emotions and cried—unlike other times. Later it turned out that he initially traveled to Cyprus.[4]

There is a well-known legend that Joseph, Mary Anna, Mary Magdalene, and several disciples were sent out to sea by the Romans in an open ship without sails, oars, or rudders; something that effectively meant a death sentence. Instead, a current would have brought the ship safely to the coast of southern France. You will find this account of the flight in *The Life of Mary Magdalene,* included in the collection of saint's lives, *The Golden Legend.*[5] This highly popular collection was written by Jacobus de Voragine and was composed between 1260 and 1263. He described the flight "in a ship without helmsman or rudder, so that they might have been drowned, but by the will of God they finally arrived in Marseille." This account was echoed in a description by Cardinal Caesar Baronius (1538–1607), curator of the Vatican Library and an important historian of the Roman Catholic Church[6] (chapter 14).

How did this story reach the Church? Prentis questioned Joseph of Arimathea about this. Joseph in regression replied: "There were many people who were loyal to me. I left instructions with those I trusted most that Mary secretly go to one of my ships. It was all well planned in advance. Do you really think I didn't know what was going to happen? Everything had to be done in great secrecy, and the sooner I left the safer it was for the followers of Yeshua. I went first to Cyprus and took another ship from there, but I knew that Mary would be safe."[7]

After this, Prentis mentioned the account of the flight as it is known

---

3. Wilson and Prentis, *The Essenes,* 206.

4. Wilson and Prentis, *The Essenes,* 214.

5. Giotto painted scenes from the life of Mary Magdalene; this fresco is entitled *The Landing at Marseilles.*

6. Cesare Baronius, *Annales Ecclesiastici,* Rome, 1595; Jowett, *The Drama of the Lost Disciples,* 33; Wilson and Prentis, *The Essenes,* 213–14.

7. Wilson and Prentis, *The Essenes,* 214.

in the Roman Catholic Church, saying: "There is a report that reached Rome that Mary, Mary Magdalene, and other followers of Yeshua were cast adrift in a boat without oars or sails."

Joseph replied: "Yes, I put that together with a Roman friend, a high-ranking official, whom I had known for many years. It was a way to keep Mary and Yeshua's closest followers safe. Once the Roman officials listed them as 'shipwrecked, presumed deceased,' they would take no further action, whatever the Sanhedrin might say. All this was planned in advance, because I knew it would be very dangerous for the whole group around Yeshua."

Prentis continued: "And did the Romans want to send that false message?"

Joseph replied: "They were open to it; I told them this would secure future supplies of tin. That was their main concern."[8]

About Joseph's statement Wilson and Prentis wrote: "So here we have a completely different scenario. It concerns a settlement about which a report will be sent to Rome. So, despite the Sanhedrin viewing Joseph as a traitor to the Jewish cause for collaborating with groups around Yeshua, the Romans support Joseph. He controls the trade in tin, a metal needed to keep the Roman war machine running. The Roman report ended up in the Vatican, where it became part of ecclesiastical historiography."[9]

Prentis asked Alariel if Joseph was visiting his sister Mary in the south of France. Alariel replied: "Yes. She lived not far from the main trade route along which the tin was transported. Convoys transporting tin regularly passed by and the Romans provided an escort of soldiers to ensure a safe passage through Gaul. There were several routes, but most tin supplies went from Marazion in Cornwall across the Channel to the port of Morlaix in Betagne. [See fig. 4.1 (map with Marazion).] It continued overland through Gaul to Limoges and Marseille. There the tin was loaded into Joseph's ships and transported to the desired

---

8. Wilson and Prentis, *The Essenes*, 214–15.
9. Wilson and Prentis, *The Essenes*, 215.

Roman ports around the Mediterranean. Joseph joined some of these convoys and spent time with his sister Mary, Mary Magdalene, and other friends and relatives in Southern Gaul. These periods were very dear to him and constituted much-needed vacations in a busy life. He looked forward to spending these periods of relative peace with the people he loved most."[10]

When Prentis asked Joseph about his relationship with England he replied (and I repeat this passage from an earlier chapter): "My tin trade took me to England a number of times. My ships made regular visits to the tin mines in the western part of that country. From my very first visit I made a point of making contact with the Druids and traveling to Avalon to meet their leaders. I discovered that they had an advanced education system with many hundreds of students in a number of sites. I maintained a good relationship with them so that if we had to flee our country, we would find a safe home here. It was necessary to plan ahead in this way, because the situation was becoming ever more difficult for the Essenes."[11]

According to Daniel Benezra, Joseph had many family members and friends in England. "It was his job to reform and revitalize the Druid way. He told them that their prophecies had been fulfilled by Yeshua because he finished a cycle for the Essenes, which Joseph did for the Druids. He finished what had basically been present for much longer. Their work was done, and they could now retire with honor. Joseph was a bridge between the Celtic and Jewish minds."

Later, Daniel spoke with Joseph in the Interlife. He heard the following from him and I summarize: The Druids believed in a creator, a preserver, and a coming savior named Yesu; many Druids were open to the teachings of Jeshua, and that helped me a lot. The love and forgiveness that Yeshua taught seemed to them a simplification and summary of all that the Druidic system contained. The teachings of Yeshua were entirely consistent with those of the Druids and were received by

10. Wilson and Prentis, *Power of the Magdalene*, 115.
11. Wilson and Prentis, *The Essenes*, 222.

many Druids as the culmination of all the knowledge they had slowly accumulated over the years. Joseph also said that not all Druids were open to the new direction and some clung to the old, but many accepted the teachings of Yeshua as the new way forward.[12]

When Prentis later asked Alariel about the prestige of the Druids in Avalon at the time of Joseph, she received the following answer:

> The Druids and their wisdom were highly respected in the West at that time. We could go so far as to say that for a cultural European of those days there were three main choices. You could go to Greece for training and sit at the feet of great philosophers; you could go to England to the Druids; or you could go to the mystery school in Egypt. These three locations offered peak experiences in education and spiritual growth. But there were subtle differences. The Greeks were more academic and intellectual; the Druids more mystical and esoteric; and the Egyptians more concerned with energies and the practical working out of esoteric ideas. So Avalon, Athens and Alexandria formed a triad of centers of education and spiritual growth. Avalon or modern Glastonbury was a major center of education and culture in Joseph's day, and its reputation as a source of wisdom and spiritual knowledge was widespread across Europe.[13]

Prentis asked Joseph what it was like to live in England after the crucifixion. Joseph responded: "My old bones didn't like the cold and the humidity, but it felt good to be more comfortable, to feel safe. The Druids helped us; we got some land that enabled us to build something and settle down."

Prentis asked, "Was that land in Avalon?"

"Yes."

---

12. Wilson and Prentis, *The Essenes*, 238–40.
13. Wilson and Prentis, *Power of the Magdalene*, 114–15.

When Prentis asked whether Joseph built a church or temple in Avalon, the emphatic answer was a very clear no. Prentis then asked, "So what did you build in Avalon?"

Joseph replied, "A very simple structure, a holy place as a place of healing and gathering."

"Were the Druids sympathetic to this?" Prentis asked.

Joseph said, "Yes, they had such places as well. Why should we build a temple; Yeshua did not ask us to build temples. . . . When the day was fine, we gathered in the open air; there were special places surrounded by great trees nobler than any temple. But when it was cold and wet—and my old bones had to endure many such days in England—we gathered in the place of healing. We knew only the simplest ceremonies and they took place at home or wherever we were. Yeshua taught us that what is important is what happens in the temple of the heart. Each of us had a temple in the heart and love is the flame on the altar there."[14]

According to an old English tradition (which will be discussed later), Joseph of Arimathea was said to have built a simple structure (later called a church) consisting of a frame of twigs covered with mud on the site where the Abbey of Glastonbury would appear later.

A SECOND TIN INTERLUDE: TIN ISLANDS IN THE NORTH

In prehistory and antiquity, Europe had few places where tin could be mined. Since the late Bronze Age in Western Europe, Cornwall and Devon in England had been the main sources of tin, particularly after the Spanish tin mines became exhausted in the third century. The Phoenicians, originally from Lebanon, traditionally traded in tin and copper starting in 1500 BCE, when they developed a trade network in which tin and copper

14. Wilson and Prentis, *The Essenes*, 224–228.

were transported from England to Marseille and from there to ports around the Mediterranean.[15]

Historical literature emphasized that the Phoenicians had ancient links with southern England through the tin trade. Phoenicia was under Egyptian influence in ancient times, and the Northern Kingdom of Israel bordered southern Phoenicia.[16] When the Northern Kingdom was conquered by the Assyrians in 722 BCE, many people from this region also fled to southern England because of the tin trade between the two areas. Two Phoenician ships from around 1400 BCE have been recovered off the coast of southern England, and numerous other archaeological finds prove the Egyptian–Phoenician influence there as well.[17]

Even before the arrival of the Phoenicians, from 2000 BCE tin was mined in western England. The old name for England was based on the tin found there; around 445 BCE, Herodotus called the area the Tin Islands or the *Cassiterides*. In addition, Diodorus Siculus (90–30 BCE) described the tin trade of that time as well as the friendliness of the local people.[18]

In the first century BCE, Diodorus Siculus wrote: "Those Britons, who live near the area that juts out into the sea or Belerion [an old name for Cornwall and the Lands' End pen-

---

15. Jowett, *The Drama of the Lost Disciples*, 42 quotes a remark by Strabo, who died in 25 CE, describing how closely the Phoenicians guarded their monopoly on the tin trade from England. When followed by other privateers, they took the wrong route and wrecked their ship rather than pointing them in the right direction.

16. van Dijk, *Maria Magdalena, de Lady van Glastonbury en Iona*, 96 states that Galilee was part of Southern Phoenicia; Jowett, *The Drama of the Lost Disciples*, 58 records the bonds of the Danites, one of the twelve Hebrew tribes with progenitor Dan with the ancient port of Tarshish at Cadiz and Massilia (Marseille).

17. Jowett, *The Drama of the Lost Disciples*, 59 mentions the trading empire of the tribe of Dan, the Danites, who traded, among others, with Tarshish, the port at Cadiz; Van Dijk, *Maria Magdalena, the Lady of Glastonbury and Iona*, 95n49 with ref. to Barry Dunford, *Vision of Albion: The Key to the Holy Grail—Jesus, Mary Magdalene and the Christ Family in the Holy Land of Britain*, 17.

18. Jowett, *The Drama of the Lost Disciples*, 42. According to Jowett, Diodorus is strongly supported by Ptolemy and Polybius; Taylor, *The Coming of the Saints*, 189.

insula], are more civilized and courteous to strangers than the rest, which is due to their intensive dealings with foreign traders. With great skill they prepare the tin that the country produces. When this precious material is first excavated, it is mixed with soil; then they separate it by melting it and purifying it. When this is done, they pour it into blocks or ingots and then transport it to an adjacent island called Ictis. For when the tide is low, the space between that island and the mainland of Britain becomes dry land; and then they carry large quantities of tin here in their carts. Here the merchants buy it and ship it to the coast of Gaul; from whence they carry it by land, with the help of packhorses, to the mouth of the Rhône in about thirty days."[19]

And elsewhere Diodorus repeated: "This tin metal is shipped from Britain to Gaul; the merchants transport it on horseback through the heart of Celtica [later France] to Massilia [Marseille] and the city called Narbo [Narbonne]."[20] (For the journey through France, see fig. 4.1 with Ictis marked on the map.)

This description explained how the tin was mined from Cornwall's mines. It was beaten into squares or ingots and taken to an island called Ictis, which was connected to the Cornish coast at low tide. From there it was shipped to the French ports on the other side of the Channel. From there, the ingots were transported on the back of packhorses over the land route across France to Marseille.[21] The existing route that Diodorus Siculus described in the first century BCE was followed a little later by Joseph of Arimathea.[22]

It seems that Joseph of Arimathea was following in the earlier footsteps of the Phoenicians. In western England, there are many traditional stories about Joseph trading with the people of

19. Diodorus Siculus, *Library of History, Volume V: Books 12.41–13,* translated by C. H. Oldfather, Cambridge, MA: Harvard University Press, 1950, 384.
20. Diodorus Siculus, *Library of History, Volume I: Books 1–2.34,* 279.
21. Diodorus Siculus, *Library of History, Book 1*; Taylor, *The Coming of the Saints,* 179.
22. Taylor, *The Coming of the Saints,* 179.

Figure 13.1. These are copper ingots from Cyprus; they weigh thirty kilograms each and have the shape of an ox hide, which makes transport easier. They date from 1500–1450 BCE and come from the "palaces" or multifunctional centers of Hagia Triada and Zakros. My photo from 2021 of ingots in the renovated Heraklion Archaeological Museum in Crete. The ingots found off the coast of Morlaix have the same shape.

Cornwall,[23] and poems and miners' songs in southern England also positively connected Joseph to the tin trade there. It was customary among the miners there to shout, "Joseph was a tin man," and "Joseph was in the tin trade."[24]

In Egypt, the divine couple Isis and Osiris were known. In Phoenicia the couple Baal and Anath were known. The people of the Northern Kingdom of Israel knew the Great Mother next to the Father. And from the Bible we know from the Northern Kingdom the Queen Mother and High Priestess Jezebel, who

23. Taylor, *The Coming of the Saints*, 180–81.
24. Jowett, *The Drama of the Lost Disciples*, 43.

took the sacred posture of the Lady in her Tower. In Western Europe, too, the agricultural population from the megalithic era, led by the Druids, had a balanced image of God. Later the Celts arrived in England and Ireland in the seventh and sixth centuries BCE. They partly adopted this egalitarian image of God; they knew the Great Mother Ana and her partner Lug.[25]

The Druids had been present in Western Europe long before the Celts arrived. It is certain that they came from a pre-Celtic culture. There was Irish lore that the first Druids came from Spain and Ireland; they appeared in the westernmost coastal regions. I wrote in an earlier book, "This raises the question whether the Druids, the educators known as the 'knowing' or 'sighted,' did not belong to an older Atlantean culture, to which the megalith builders also belong."[26]

Several authors have pointed out that there has been a connection between the house of David and the royal houses of Britannia or Britain.[27] Barry Dunford points to an ancient tradition that Anna, Mary's mother, was originally from the Cornish royal house. According to this tradition, Joseph of Arimathea was Anna's brother and a member of the Cornish royal family.[28] Earlier it was mentioned that Joseph was Anna's son.

A CELTIC INTERLUDE:
DRUIDS, YESHUA, AND JOSEPH IN BRITAIN

The relationship of the family of Yeshua to England was previously discussed in a nutshell in An English Interlude. Let me summarize: In 1961, George F. Jowett wrote the book

25. Jones, *In the Nature of Avalon*, 147: The Celts had a less patriarchal religion than today's religions.

26. van der Meer, *The Black Madonna from Primal to Final Times*, 197n92.

27. Capt, Jowett, Mock and Elders. See van Dijk, *Maria Magdalena, de lady van Glastonbury and Iona*, 106.

28. van Dijk, *Maria Magdalena, de lady van Glastonbury and Iona*, 107.

The Drama of the Lost Disciples, which was based on historical sources and legends. It was so popular in England that by 1993 it was reprinted in twelve editions and was later published in new editions. Based on the work of previous researchers, Jowett emphasized that in the traditions of Cornwall, Devon, Somerset, Wiltshire, and Wales it was an established fact that Joseph of Arimathea visited Britannia in the company of the young Yeshua.[29] But legends are not hard evidence. John Michell argued that even the skeptical historians in England were convinced of the antiquity and importance of the legend that Yeshua visited Britannia and that Joseph of Arimathea started a community here, but that it can never be proven with precise details.[30] In English literature the following was emphasized: Joseph of Arimathea was the apostle of England and the founder of the early Church in England.

Ancient writers report positively about the Druids. Diodorus Siculus wrote that some sing and accompany themselves on the harp. Pomponius Mela wrote about how to understand the movements of the stars. Strabo stated that they study the natural sciences and ethics and teach about the immortality of the soul.[31] And Julius Caesar also reported in *De Bello Gallico* about the many young people who study with them. He wrote about verses to be memorized, the immortality of the soul and that the soul moves from one body to another, the movements of stars and constellations, the transmigration of souls after death, the size of the world and countries, and about nature and the essence of things. It took twenty years to go through the entire Druidic curriculum. Subjects taught included astronomy, medicine, poetry, law, natural philosophy, public speaking, and the teaching that souls are immortal

---

29. Jowett, *The Drama of the Lost Disciples*, 69.
30. Michell, *New Light on the Ancient Mystery of Glastonbury*, 82.
31. Taylor, *The Coming of the Saints*, 189.

and pass from one body to another after death. At the time of Yeshua there were forty Druidic universities in England. Some sixty thousand students were enrolled, including children of the English nobility and children of the elite of the European continent.[32] John Michell—supported by many others—did not rule out that Yeshua was taught by the Druids during his stay in England.[33] He argued that there was evidence that Avalon/Glastonbury was the center of Druids and Celtic Christianity.[34]

George Jowett argued that the name Yesu has been known to the Druids long before the arrival of Jesus,[35] and the idea that the Druids were aware of Yesu as the bearer of light is supported from an esoteric angle. The founder of anthroposophy, Rudolf Steiner, stated that the Celtic Druids clairvoyantly perceived the events surrounding the crucifixion in Jerusalem. The Druids had been venerating the Cosmic Sun for many centuries.[36] The Druids in Wales, southern England, and Ireland had been expecting the solar logos for many years. They knew of the coming of Yeshua as bearer of the solar logos to Earth. They regarded him as the Sun Being who would bring cosmic wisdom and cosmic love to Earth. Anthroposophist Danielle van Dijk stated "The Sun Being was together with his beloved goddess, and therefore the Druids could easily accept Mary Magdalene as the incarnation of the great goddess."[37]

---

32. Jowett, *The Drama of the Lost Disciples*, 137n1; Prophet, *Mary Magdalene and the Divine Feminine*, 253.

33. Michell, *New Light on the Ancient Mystery of Glastonbury*, 81.

34. Michell, *The New View over Atlantis*, 167.

35. Jowett, *The Drama of the Lost Disciples*, 51, 78 discusses Yesu as part of the Celtic Druidic trinity.

36. Jakob Streit, *Sonne und Kreuz*, Stuttgart, 1977; Dutch transl.: *Zon en Kruis. Van steencirkel tot vroeg-christelijk kruis in Ierland*, Rotterdam: Christofoor, 1980, 69 with ref. to Steiner.

37. van Dijk, *Maria Magdalena, de lady van Glastonbury and Iona*, 115, 128, 161n71 and 165n74, 166: Jesus as carrier of the great Solar-logos; Hartley, *The Western Mystery Tradition*, 32.

Jowett pointed out that because of the presence of Yeshua, Joseph, and Mary Magdalene, the ancient Celtic religion in Britannia and Gaul was slowly beginning to transform into the Way.[38] From the French point of view, even if there are no historical sources supporting it, the only conclusion is that Druid leaders supported the arrival of Christianity and that the people followed their footsteps. The rapid spread of Christianity can only be explained if it took place with the consent of the Druids.[39] It had been a harmonious transition and a peaceful process, contrary to what was later described in Roman historiography about allegedly barbarian Druids performing human sacrifices. It's the old saying again: "The victors write the history, and the losers take the brunt." After the conquest by the Romans of England in 44 CE, many Druids were said to have fled to free Ireland.[40]

After the arrival of Yeshua, Joseph of Arimathea, and his companions, Mary the mother and Mary Magdalene in and around Glastonbury (you will find their traces in the landscape, among other places) a synthesis developed between the Druidic wisdom and the Hebrew knowledge of the people of the Way. Hebrew Christianity was cosmically oriented; it knew the knowledge of the heart and had an image of God where feminine and masculine were still in balance. This was brought to Ireland, England, and France and there, in synthesis with the cosmic and natural religion of the Druids and Celts, developed into Celtic Christianity.

My tutor, Professor Gilles Quispel, convinced me of the connection between ancient Hebrew or Aramaic Christianity and Celtic Christianity. Celtic Christianity is seen as the last

---

38. Jowett, *The Drama of the Lost Disciples*, 62, 78, The merging of the British Druidic Church with Christianity was a normal procedure, peacefully performed.
39. Alexandre, Bertrand, *La Religion des Gaulois*, Paris: Leroux, 1887; Jakob Streit, *Zon en Kruis*, n.p.: n.p., 69n97.
40. van Dijk, *Maria Magdalena, de lady van Glastonbury and Iona*, 162, 165.

flourishing representative of Aramean, Syrian, and Egyptian Christianity.[41] In the early period it had an egalitarian image of God; both female and male clergy were highly respected in that early period when Roman Christianity had not yet penetrated Ireland and England.

In Avalon: The Legends and Historical Sources

According to George Jowett, Joseph traveled to Britannia for several reasons prior to the crucifixion. He traveled on business related to tin deliveries. In addition, because of his noble Celtic descent, he visited his family from his mother's side. But the third motive was contact with the Druids. Jowett stated that news of the crucifixion was known to the Druids in England even before Joseph landed in France; the news had preceded him.[42] A Druidic delegation had traveled to Gaul to meet Joseph there and invite him to Britannia.[43] According to Jowett, "They knew and wanted Joseph to be the head of the British Church. He was said to have arrived in AD 36 and built the first 'old church' between AD 36 and 37."[44]

Significant is the wall painting of Joseph in St. Patrick's Chapel in the grounds of Glastonbury Abbey (fig. 13.0), which shows Joseph with a floppy hat and traveling staff arriving in Avalon, the island on which Glastonbury later appears. To his left you see his ship on a wide river with a clear hill further in

---

41. Roger E. Reynolds, "Virgines Subintroductae in Celtic Christianity," in *Harvard Theological Review* 61 (1968), Cambridge: Cambridge University Press, 547–566, 552n30, 556, published online by Cambridge University Press in 2011.

42. Jowett, *The Drama of the Lost Disciples*, 61, 68, 73 states that Joseph went back to England at the invitation of Prince Arvigarus, who offered him land, security, and protection from the Romans; he is cousin of Caratacus. He and his father Cunobelinus formed the royal dynasty of the Silures

43. van Dijk, *Maria Magdalena, de lady van Glastonbury and Iona*, 128n60.

44. Jowett, *The Drama of the Lost Disciples*, 82 follows the account of the historian Gildas in *De Excidio Britanniae*, who mentions that this was completed in the last year of the reign of Emperor Tiberius in 37.

the background; he walks on a plank to the shore and plants his staff in the earth. It blooms immediately. This is how the legend was depicted.

Joseph was welcomed by King Arviragus of Siluria from the first century. According to the Hardyng Chronicle, a fifteenth-century document that goes back to much older documents, the king donated Joseph and his eleven disciples twelve hides of land (1,920 acres) on the island of Avalon, a small six square kilometers of land. No tax had to be paid on this donation—a privilege that, according to the Dutch author Danielle van Dijk, only very important Druids enjoyed at that time. In addition, they were given an absolute safe conduct to move from village to village. They had the status of immunity. No government servant from outside the twelve hides area, including the king, had authority over this area, so it was actually a kind of free state. This royal gift was confirmed in *The Domesday Book*, which was compiled in 1086 by order of the Norman King William the Conqueror to levy taxes in the territory he conquered: "The Domus Dei: in the great monastery of Glastonbury, called The Secret of the Lord; This Glastonbury Church owns on its own land XII (12) hides of land on which no tax has ever been paid." This donation is a historical fact.[45]

In fact, the king bestowed Joseph with the most sacred ground he had. Why? Glastonbury, or the Isle of Glass, was a very sacred place in ancient times. In myth and fairy tale, a place of glass is a place of enchantment, a doorway to the Other World. In Britannia it was believed that here in Avalon the veil to the Other World was thin and permeable. From there, according to Danielle van Dijk, the dead have easier access to the afterlife and souls can be reborn more quickly on Earth.[46]

---

45. Jowett, *The Drama of the Lost Disciples*, 144n1 with ref. to *Domedsay Survey Folio*, p. 249B

46. van Dijk, *Maria Magdalena, de lady van Glastonbury and Iona*, 116.

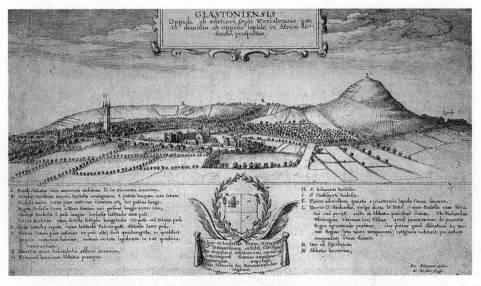

Figure 13.2. Wenceslas Hollar's engraving of the precinct at Glastonbury Abbey in the mid-17th century.

The Island of Glass is also called Avalon or The Apple Island; the Celtic word for "apple" is *aval*. It is the sacred fruit of the Druids, a symbol of fertility. Avalon is a kind of paradise and is also called Paradise. The apples refer to the fruits of the Tree of Life. Life is abundant: the land is fertile and provides food in abundance. In addition, the island is surrounded by fish-rich waters of the River Brue. In short: Glastonbury was holy ground.

The Apple Island is the seventh and highest island among six smaller ones. The island was shaped like a swan in flight (fig. 13.3). The swan had a round body and a long neck and flew from northeast to southwest.[47] Avalon consisted of four hills: the towering Tor is in the center. Close by was the round Chalice Hill with two springs: from one spring flowed white water and from the other red water; it was ferruginous water. St. Edmunds Hill and Wearyall Hill lay to the southwest in

---

47. Jones, *In the Nature of Avalon*, 14.

the long neck of the flying swan. During the Ice Age, the plains of Somerset were dry. But around 10,000 BCE the ice melted and the sea level rose. Around 5000 BCE the four-hill area became an island that was cut off from the mainland at high tide. Rivers supplied fresh water and a swampy area was created; people laid wooden planks over the swampy ground to reach the seven islands.[48] The rivers constantly changed their course but brought a lot of fresh water into the area.

Between 1000 BCE and 300 CE large seagoing vessels could

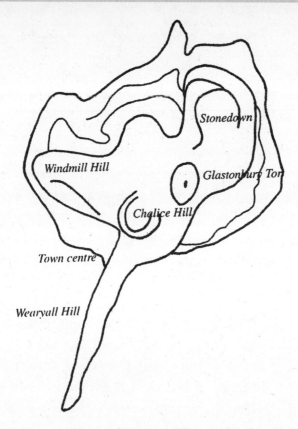

Figure 13.3. The island of Avalon resembles a swan in high flight.
From Kathy Jones, *The Nature of Avalon*, 20
(for a more detailed map, see 107).

---

48. Michell, *New Light on the Ancient Mystery of Glastonbury*, 29.

sail up the rivers Axe and Brue. When you sailed more than nineteen kilometers up the lower reaches of the River Brue, you came to a cluster of islands. One was the Isle of Avalon, with the *Inis Wytren* or Glass Island less than a mile away; this was because glass was made here and the waters around it would be crystal clear.[49] Hence, also the name of the town of Glastonbury that was situated on the island of Avalon. The area had been inhabited since Neolithic times, and by around 300 BCE the village of Glastonbury was known. There was evidence of seagoing vessels coming to the ports to fetch silver and lead from the Mendip Hills north of Glastonbury. The Romans and the later monks of Glastonbury Abbey carried out much drainage work in the plain and it became dry. Today, much of the plain lies slightly below sea level. Continuous pumping and drainage are required to protect the plain from salinization and flooding.[50]

About the year 35 CE (some say the year 31),[51] Joseph of Arimathea went to Britannia, sailed his ship nineteen kilometers up the River Brue, and reached the island of Avalon. The travelers, according to tradition, climbed the first hill at the extreme point of the long swan's neck. Legend has it that Joseph, tired from the journey, stopped to rest. He planted his staff in the ground and immediately the staff began to bloom, an ancient theme that occurs more often in the Bible at very important events that were blessed. Joseph was weary or extremely tired from climbing the hill,[52] therefore, the place where the staff was planted was called Weary All Hill; today it is called Wearyall Hill. It was formerly accessible only by ship. According to Kathy Jones, who has lived in the area for more

49. Jowett, *The Drama of the Lost Disciples*, 74–75.
50. Mann, *The Isle of Avalon* 11.
51. Taylor, *The Coming of the Saints*, 195, gives the information on an old plaque in octagonal shape on the abbey grounds.
52. Jones, *In the Nature of Avalon*, 84.

than fifty years, it was the first natural landing place for travelers approaching Glastonbury by water as they sailed up the River Brue from the sea.

On the hill where the staff took root in the earth and bloomed, the staff transformed into what is called the blooming Holy Thornbush or the Holy Thorn. Thorns often appeared in places where the Earth's energy is high. It was a hawthorn that bloomed with bright white blossoms. The special thing was that this hawthorn bloomed twice a year, exactly as in Galilee—once in the spring and once in the winter—while the European hawthorn only blooms once a year. The current hawthorn is a cutting of a cutting of a cutting. It stands on top of the hill and is bent in a northeasterly direction by the blowing southwesterly winds.[53] The Holy Thorn blooms in spring but also from late November to January; it is in full bloom at the beginning of January. According to Jowett, it was a living reminder of the birth of Yeshua in Bethlehem and the landing of his followers in Britannia.[54]

The mural depicting Joseph's landing in St. Patrick's Chapel on the grounds of Glastonbury Abbey has already been discussed. Joseph with a floppy hat and traveling staff arrived in Avalon: his ship and the spacious river with a clear hill further in the background are depicted; he walked across a jetty onto the shore and planted his staff in the earth. It immediately bloomed as the Holy Thorn. This was how the legend was depicted in the image. It was not unimportant that he held something special in his hands. The so-called wattle church, or Old Church, was round and built of willow branches and mud, as was customary at that time. When I was there in 2017, I took a picture of it[55] (fig. 13.0).

---

53. Jones, *In the Nature of Avalon*, 84–86; van Dijk, *Maria Magdalena: de lady van Glastonbury and Iona*, 129.
54. Jowett, *The Drama of the Lost Disciples*, 74.
55. van Dijk, *Maria Magdalena, de lady van Glastonbury and Iona*, 149.

Figure 13.4. The sign indicates the location of the wattle church or Old Church, with the Lady's Chapel on the right and the source on the south side.

In the grounds of Glastonbury Abbey there is a pit with the sign "Wattle Church" next to the large ruin of the abbey church (fig. 13.4). It would have been the very first church, a building with a framework of willow branches covered with dried mud. The structure was called the Old Church or in Latin the *Vetusta Ecclesia*.[56] According to tradition, Joseph and his group built the first "church" in 37 CE or possibly even earlier.[57]

The archaeologist and anthropologist E. Raymond Capt wrote in 1983 that the pilgrims "erected a round shape what must have been the first Christian church." Studies by the Somerset Archaeological Society show that the first building was round and had a diameter of 7.62 meters. There are twelve huts

---

56. Taylor, *The Coming of the Saints*, 195 gives a measurement of sixty feet long by twenty-six feet wide (18.2m long by 7.90m wide); Jowett, *The Drama of the Lost Disciples*, 76 gives the same information; Prophet, *Mary Magdalene and the Divine Feminine*, 239 does likewise.

57. Taylor, *The Coming of the Saints*, 195.

Figure 13.5. Old Church or wattle church of willow and mud;
reconstruction by Frederick Bligh Blond with the "church" in
the center and the twelve residential units around it.
From John Michell, *New Light on the Ancient Mystery
of Glastonbury*, 81, 86.

in a circle around it. Reconstruction drawings exist[58] (fig. 13.5). The central round building is dedicated to Mary.

To protect the round building, it was framed with wood in 630 as a kind of relic and fitted with a lead roof. The Old Church dedicated to Mary stands in a chapel dedicated to Saint Mary. Later this sacred space became the chapel of St. Joseph, as mentioned on the plan (fig. 13.6).[59] When William of Malmesbury arrived at Glastonbury Abbey in 1129, the Old Church was still standing in the Marian Chapel; he wrote about it in awe. But in 1184 one of the tapestries in the Old Church, which was dimly lit by candles and an oil lamp, caught

---

58. Michell, *New Light on the Ancient Mystery of Glastonbury*, 80; Michell, *New View over Atlantis*, 175 gives the reconstruction of Bligh Bond.
59. Jowett, *The Drama of the Lost Disciples*, 76; Taylor, *The Coming of the Saints*, 195. The ground plan is on 192–93.

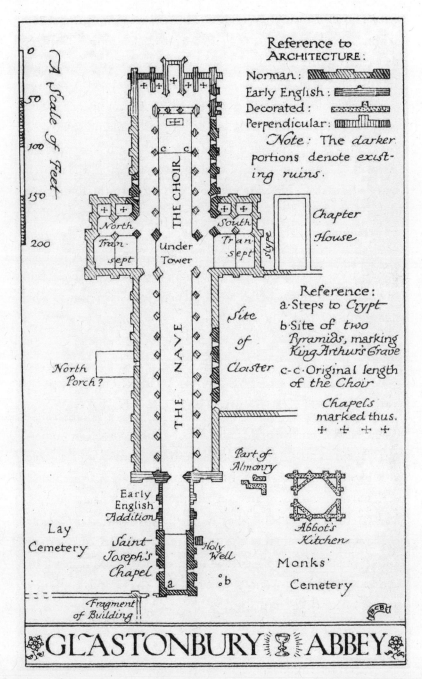

Figure 13.6. Floor plan of Glastonbury Abbey showing St. Joseph's Chapel, originally called Mary Chapel, at the bottom, with the Holy Well to the right of the side wall. From John Taylor, *The Coming of the Saints*, 102–103.

fire. The Old Church, the abbey church, and the entire abbey with its library and cloister, were burnt to the ground. William was the important witness who saw the Old Church before it finally burned down.

The visible evidence of Avalon's early Christian history is now gone forever. But the Old Church was and remains connected through legends and church sources with Yeshua, Joseph of Arimathea, and Mary, the mother of Jesus, to whom the church was dedicated. Celtic churches were dedicated to the founder of the church, and so the Reverend Lionel Smithett Lewis, vicar at Glastonbury, in his 1922 scholarly book, *St. Joseph of Arimathea at Glastonbury*, raised the possibility that Mary the mother may have been at Glastonbury and may have been buried there.[60] But there are also traces of Mary Magdalene in Glastonbury. The question is: Which Mary was it?

AN ECCLESIASTICAL INTERLUDE:
JOSEPH OF ARIMATHEA IN ENGLAND

There is an extensive oral tradition that was recorded in numerous legends and songs (especially miners' songs) over a large area in southwest England. In addition, there are historical and ecclesiastical sources that make the presence of Yeshua, Joseph, and the Marys plausible, but not fully provable. I list below the sources that support the antiquity of the British Church and Yeshua and Joseph's part in it, in chronological order. The ecclesiastical sources seem to confirm that the legends contain a grain of historical truth.

• The church fathers Tertullian (160–230) and Origen (182–253) referred to the old British church around 200 CE.

---

60. Michell, *New Light on the Ancient Mystery of Glastonbury*, 81.

- Critics argued against the antiquity and primacy of the British Church as follows: Bishop Irenaeus of Lyons (130–202) listed all Christian churches but made no mention of the Celtic Church in England. From a Roman Catholic point of view, the Church in England was a heretical and half-pagan institution because of its connection with the Druids and Celts. But according to the advocates who took the legends seriously, this does not mean that the Celtic church did not exist in Irenaeus's time. He ignored the existence of the Celtic Church because of the deviation from Rome. Only churches that were oriented toward Rome counted for him.

- From the early third and fourth centuries there were Eusebius, Bishop of Caesarea (260–340) and Hilary of Poitiers (300–367). Both had written that some of the apostles left Palestine immediately after the crucifixion and went to Glastonbury after landing in southern France. This has been confirmed by many historians in the following centuries.

- Three British bishops were present at the Council of Arles in 314; at subsequent councils it was customary to give precedence to British bishops on account of the fact that the first Christian church was founded in Glastonbury.[61]

- Maelgwyn of Avalon wrote around 450 that "Joseph of Arimathea, the noble decurion with his eleven companions, found his final resting place on the island of Avalon." Many later historians referred to this statement.[62]

---

61. Taylor, *The Coming of the Saints*, 314; Michell, *New Light on the Ancient Mystery of Glastonbury*, 77.
62. Jowett, *The Drama of the Lost Disciples*, 229.

- The honored historian Gildas the Wise (500–570), who lived in Glastonbury for a time around 520, was widely and well-respected until modern times. He had access to all original documents in the abbey. On this basis he mentioned in *On the Destruction of Britain* (or in Latin *De Excidio (et conquestu) Britanniae*) that Joseph of Arimathea came to England and lived and died there. He gave a precise date for Joseph's arrival. He wrote that Joseph brought Christianity to England and came to Glastonbury "in the last year of Emperor Tiberius's reign." Tiberius reigned for twenty-two years; his last year was 37 CE.[63] The year 37 was only a few years after the crucifixion; sources and legends emphasized that Joseph arrived "in the years immediately after the crucifixion."

- A further testimony was given by the Briton Augustine of Canterbury (circa 530–604). He ended up in Rome and was sent back by the Pope of Rome to spread the Roman Catholic faith in England. In 597 Augustine wrote a long letter to the Pope from England, in which he confirmed the presence of the young Jesus in England and stated that "Christ himself built a church here and dedicated it to his mother." Augustine of Canterbury was a Roman Catholic clergyman; he was regarded as the apostle of the Anglo-Saxons and the founder of the Church of England—despite the fact that he testified that it was much older.[64] Rightly, Jowett, supported by numerous others, noted that this conventional view is an error.[65]

---

63. Taylor, *The Coming of the Saints*, 176, 191, and 202 with reference to Gildas, *History of Gildas*, chap. 8 and 9; Jowett, *The Drama of the Lost Disciples*, 82, 230.

64. Augustine, *Epistolae ad Gregorium Papam*; Beda, *Hist. Eccl.* book 1 chap. 30 contains message of Pope Gregory to Augustine of Canterbury; Jowett, *The Drama of the Lost Disciples*, 138.

65. Jowett, *The Drama of the Lost Disciples*, 79.

- Finally, there was William of Malmesbury who in 1129 was invited by the Abbot of Glastonbury Abbey to write a reliable history about the abbey based on ancient documents: *On the Antiquity of the Church of Glastonbury* (or in Latin *De Antiquitate Glastoniensis Ecclesiae*). William wrote in awe of the wooden church, its powerful appearance, and the altar that was overflowing with relics of saints from top to bottom and also inside.[66] In the original version of his work, William cited early second-century sources about the Church of Glastonbury, but he also found references in even older documents to an earlier foundation. He wrote "the church at Glastonbury was the first and oldest church in England" and "the church was made by God and dedicated to himself and to Mary, the holy mother of God."[67] In thirteenth-century copies of William's work, to which monks later added all kinds of things, it was stated that the disciples, led by Joseph of Arimathea, arrived in England in 63 CE and that they were told in a dream by the angel Gabriel to build a church dedicated to the holy mother of God, the Virgin Mary.[68] This version seems inconsistent with older twelfth-century accounts both from William himself and from many illustrious historians before him. Joseph was there much earlier and much more frequently.

There were very early reports in England of the presence of young Yeshua accompanied on his journey north by his uncle Joseph of Arimathea. There are reports of the presence of Yeshua as a young man and Mary his mother. After

---

66. Michell, *New Light on the Ancient Mystery of Glastonbury*, 86–87, 90.
67. Michell, *New Light on the Ancient Mystery of Glastonbury*, 90.
68. Michell, *New Light on the Ancient Mystery of Glastonbury*, 91–92; Jowett, *The Drama of the Lost Disciples*, 230.

the crucifixion, there are reports of the settlement of Joseph of Arimathea with eleven companions in Avalon. And there are reports of the presence of Mary Magdalene. There is more information about this later. When you review all the historical and ecclesiastical source material in combination with the legends and oral tradition, it seems very likely that the reports of the presence of Yeshua, Joseph, and the two Marys can be taken seriously.

IN THE FOOTSTEPS OF MARY MAGDALENE

Figure 14.0. Giotto di Bondone, *Scenes from the Life of Mary Magdalene, Her Journey to Marseilles*, 1320, Lower Church of the Basilica of Saint Francis, Assisi.

14
MARY MAGDALENE
IN FRANCE

Landing in France:
The Regression Reports

When Prentis asked Daniel Benezra what happened to Mary (the mother of Yeshua), Martha, and Mary Magdalene after the crucifixion, he said, "I heard that a short time after Joseph was to leave, these three sailed across the central sea to reach the Celtic lands. But I never heard any further reports of them. The movements of the core group were known only to a few. Much confusion was deliberately spread to cover their tracks, and there were many conflicting rumors . . . I did not learn until much later that Joseph reached England safely."[1]

Prentis asked Joseph of Arimathea what happened to Mary (his sister) after he had left in a hurry. I repeat an earlier comment by Joseph: "There were many who remained faithful to me. I left instructions with those I trusted most so that Mary could secretly leave on one of my ships. It was all planned in advance. Did you really think I didn't know what was going to happen? Everything had to be done in great secrecy and the sooner I left the safer it would be for the followers of Yeshua. . . . I knew Mary would be safe."[2]

---

1. Wilson and Prentis, *The Essenes*, 204.
2. Wilson and Prentis, *The Essenes*, 214.

Prentis asked if Joseph knew the story that reached Rome, that Mary and Mary Magdalene and other followers of Yeshua were forced to embark in a boat without oars or sails and set adrift. Joseph replied to this—and I repeat—that he had invented this ruse himself together with his Roman friend, a high-ranking official. "It was a way to protect Mary and Yeshua's closest followers. Once the Roman officials had listed them as 'castaway, presumed dead,' they would take no further action, whatever the Sanhedrin might say. All this was planned ahead, for I knew it might become very dangerous for the whole group around Jesus."[3]

To summarize: Immediately after the crucifixion, Mary Magdalene also left Palestine on board one of Joseph of Arimathea's ships and went to Alexandria. After a while, the extended family group gathered there and sailed to southern France, then known as Gaul. There Mary Magdalene and her group received a lot of help and support from the Jewish community.

The Immediate Family of Mary Magdalene in the South of France: The Regression Reports

Prentis asked: "Did she have any close relatives waiting for her in France?"

Alariel replied, "Yes, Mary Anna's brother had moved to the Languedoc with his wife Tabitha. Their daughter Sarah, a disciple, and her husband Philip had joined them. By that time, Isaac's brother Jacob had also moved to the Languedoc, so there was a complete family network there to support Mary."

Prentis asked whether the arrival of all these advanced people raised the vibration levels of the Languedoc.

Alariel replied, "Yes, very much. Such a gathering of advanced initiates with Mary Magdalene, who focused the energy of the Sacred Feminine, led to the development of an advanced and subtle spiritual culture in that realm. The power of the Sacred Feminine increased in

---

3. Wilson and Prentis, *The Essenes*, 215.

the hearts and minds of the people there, culminating in the Cathar impulse about 1,200 years after the arrival of Mary Magdalene. And it all started with the arrival of Mary and her extended family group. There was already an established Jewish community in Languedoc when they arrived, so the way was smoothed for them."[4]

Sarah Anna, the Daughter of Mary Magdalene and Yeshua: The Regression Reports

Mary Magdalene boarded the ship that took her to France with her three-year-old daughter. Prentis wanted to know more about that and asked Alariel if Sarah was the name of Mary Magdalene's daughter.

Alariel responded, "Yes and no. Sarah was more of a title than a name, a title acknowledging her special lineage. Her real name was Anna; she was named in honor of Yeshua's grandmother Anna."[5]

Prentis: "How could Sarah be three years old at the crucifixion when Yeshua and Mary were married during the public years?"

Alariel replied that the marriage in Cana took place five years before the crucifixion.[6]

Prentis then asked, "Did a bloodline extend from Sarah Anna?"

"Yes, reaching down to the present day. Sarah Anna is the unknown princess of the Western world, just as Mary Magdalene is the unknown queen. They were not concerned about being known, because their kingdom was not of this world."

Prentis said, "So Yeshua and Mary didn't have children with the intention of starting a dynasty?"

"Absolutely not. They were working for the kingdom of heaven and

---

4. Wilson and Prentis, *Power of the Magdalene*, 129-130.
5. Wilson and Prentis, *Power of the Magdalene*, 130. This is confirmed by a remark of Laurence Gardner, *The Magdalene Legacy*, 31–32, where it is stated that Sarah means "princess."
6. Wilson and Prentis, *Power of the Magdalene*, 130; Claire Heartsong gives the year 24 as the date.

had no interest in planting descendants on the throne of Israel or on any other throne."[7]

Prentis: "If was not about their descendants forming a dynasty, then what was the real purpose?"

Alariel replied that they provided a constant counterbalance to the patriarchal energy that has come to dominate the Western world. The descendants of Mary Magdalene carried the energy of the Sacred Feminine, with all its potential for balance and healing. The Magdalene lineage has provided a suitable channel through which many priestesses of Isis could incarnate, so the Isis vibrational frequency could continue to serve and bless the Earth. From our point of view, this is much more important and beneficial for humanity than any dynasty of kings.

"And there was only this one daughter?"

"Yes, but the situation was a bit more complicated than most people realize. Advanced creatures like Yeshua and Mary Magdalene had options not available to the average person. Sarah Anna was light conceived."[8]

The Merovingian Kings Descended from the Three Other Children: The Regression Reports

Prentis said to Alariel, "There has been a lot of speculation about the bloodline of Mary Magdalene. Some authors mention three children, two boys and a girl named Tamar."

Alariel explained that Mary Magdalene adopted three other children prior to the crucifixion, partly as a kind of cover and protection for Sarah Anna and her descendants:

During the period that Mary Magdalene lived in the Languedoc, she lived here with her daughter Sarah Anna and three other chil-

7. Wilson and Prentis, *Power of the Magdalene*, 131.
8. Wilson and Prentis, *Power of the Magdalene*, 131–32.

dren whose father had died; they were cousins of Yeshua: the eldest was called Tamar Miryam and there were two boys Yeshua and Josephes, also called Joses. Mary adopted them before the crucifixion and adoption was a common practice in those days of high adult mortality. These three children had, at the soul level, volunteered to provide a smokescreen, a shield for the true bloodline represented by Sarah Anna.[9]

It was a great privilege and a big spiritual opportunity to spend time in a family headed by an advanced being such as Mary Magdalene and despite the risks involved, there was no shortage of volunteers at the soul level for this task. It was the descendants of these adopted children who eventually became the Merovingian kings. As children of the house of David, they certainly had royal blood, but their bloodline was not quite as special as some people may have supposed.

Prentis then said, "Much of the modern speculation about the bloodline focuses on whether there are now descendants of Yeshua and Mary still walking the earth and what might happen if they made themselves known."

Alariel replied that:

the bloodline is no longer relevant in today's modern democracies; the times when countries were ruled by priest-kings is over and it would put back the clock. In contrast, Mary Magdalene's position as the spiritual partner of Yeshua and her role as an empowered and enlightened woman is relevant today and we suggest that you focus on that. The bloodline through Sarah Anna continues to this day. But . . . it is not about a bloodline but about the vibration that emanated from the Isis frequency. This allowed priestesses of Isis to reincarnate and counterbalance the patriarchalization of Western society.[10]

9. Wilson and Prentis, *Power of the Magdalene*, 132.
10. Wilson and Prentis, *Power of the Magdalene*, 131–33.

THE ARRIVAL INTERLUDE:
THE HISTORICAL FACTS BEHIND
MARY MAGDALENE'S LANDING IN FRANCE

There are numerous traces that made the presence of Mary Magdalene in the south of France plausible. First, there were numerous canonical texts describing the landing of the group from Palestine. In the Parisian libraries there are authentic texts on this subject dating back to the second and fourth centuries. The historical and ecclesiastical reports of their presence in Provence are old and confirm the flight of Mary Magdalene and her retinue to France.

Rabanus Maurus

Maurus was a Benedictine scholar, the Archbishop of Mainz and Abbot of the Fulda Monastery, the most important educational center under Charlemagne. He was known as the most eminent scholar of his time; he had no equal where Bible knowledge was concerned. His main work, *The Life of Mary Magdalene,* consisted of fifty chapters or books, bound in six volumes, which were beautifully illustrated.[11] It included much material from the fourth century. It began like this: "The contemplative life of the blessed Mary Magdalene, with deep deference named as the sweetest chosen of Christ and the beloved of Christ." Here you can see how the early Church respected and appreciated Mary Magdalene.

Further on in chapter 37 about the landing it follows: "Entrusting himself to the waves of the sea, St. Maximinus archbishop—together with the glorious friend of God Mary Magdalene, her sister blessed Martha and blessed archbishop

---

11. Jacques-Paul Migne, *Patrologia Latina,* n.p.: n.p., parts cvii–cxii; David Mycoff, *The Life of St. Mary Magdalene and of Her Sister Saint Martha,* Kalamazoo, MI: Liturgical Press, 1989.

Parmenas and the bishops Trophimus and Eutropius, along with the rest of the leaders of the army of Christ—left the coast of Asia Minor behind him. And thanks to a favorable east wind they crossed the sea between Europe and Africa, all the while keeping the city of Rome and all of Italy to their right. . . . They then changed course to the right and reached the city of Marseille in the Gaulish province of Vienne, where the River Rhône flows into the sea."

This passage confirms the landing in Marseille. They traveled in an ordinary ship in which they determined the course themselves. "There, calling on God the great king of all the world, they divided the provinces among themselves, where the Holy Spirit had led them. . . . And soon they separated and preached everywhere and the Lord worked with them and confirmed their words with subsequent signs. To the archbishop Saint Maximinus fell by fate Aix (Aquensem) . . . where also blessed Mary Magdalene ended her wanderings."[12]

Immediately after that follows a long list where the groups of disciples went. *The Life of Mary Magdalene* was important because in chapter 37 it described in detail the dispersal of the people of the Way across France and Spain.

- Seventeen disciples went to seventeen different capitals of regions in France. One of the seventeen disciples is Parmenas, who went to Avignon and took with him Martha, her maidservant Marcella, Epaphras, Sosthenes, Germanus, Euchodia and Syntex.
- There were seven disciples who went to seven regions in Spain.[13]

12. Mycoff, *The Life of St. Mary Magdalene*, chap. 37, 93; Taylor, *The Coming of the Saints*, chap. 5 "The Story of Rabanus," 92–121.
13. Mycoff, *The Life of St. Mary Magdalene*, chap. 37, pp. 94–95, lines 2165–2205; Taylor, *The Coming of the Saints*, chap. 5 "The Story of Rabanus," 105–07; Gregory of Tours stated that seven bishops were sent to Gaul.

- In chapter 38 there is information about Maximinus who went to Aix together with Mary Magdalene.
- In chapter 39 *it was denied* that Mary Magdalene lived as a hermit for many years.[14]
- In subsequent chapters you will read a lot about Martha.
- In chapter 45 the death of Mary Magdalene is explained, and there you will find information about her tomb, which was tended by Maximinus.
- In chapter 47, after her death, Mary appeared to her sister Martha with a flaming torch in her hand. With this she lit seven candles and three lamps that had gone out. It was possible that Mary Magdalene was therefore remembered in France as *La Dame de la Lumière* or the Lady of the Light.

Jacobus de Voragine

The medieval author Jacobus de Voragine (1229–1298) was archbishop of Genoa. According to Church archives he wrote *The Legend of Saint Mary Magdalene* between 1260 and 1264. Jacobus de Voragine claimed to derive his information from ancient sources such as the second-century Church Father Hegesippus and the first-century ancient author Flavius Josephus. However, this can no longer be traced.[15]

According to de Voragine, Mary Magdalene was of "noble descent with parents descended from kings." The text later in part 1 reads:

In the 14th year after the Passion of the Lord and his Ascension into heaven (AD 44), long after the Jews had killed Stephen and banished the other disciples from Judea,

14. Mycoff, *The Life of St. Mary Magdalene*, chap. 39, p. 98, lines 2315–22.
15. Beavis and Kateusz, *Rediscovering the Marys*, 232.

the disciples departed for the lands of different nations and there spread the word of God. In that time one of the 72 disciples, the blessed Maximinus, was with the apostles. It was Maximinus to whose care Peter had entrusted Mary Magdalene. During the diaspora, Maximinus, Mary Magdalene, Lazarus, her sister Martha, Martha's handmaid Marthilla, the blessed Cedonius, who was born blind, and many other Christians, had been put to sea in a ship without a helmsman or rudder by unbelievers, so that they could have drowned, but through God's will they eventually came to Marseille.[16]

Here, again the landing in Marseille was confirmed, only the group reached this port in a ship without a helmsman or rudder.

Cardinal Baronius

The information that Jaccobus de Voragine included in the biography of Mary Magdalene between 1260–1264 was largely repeated in the sixteenth century by Cardinal Baronius. He was a historian and librarian of the Vatican; he cited unnamed sources, but probably drew on the archives of the Vatican. "In that year the company of Joseph of Arimathea and those who went into exile with him were put to sea in a ship without sail or oars. This ship drifted and finally reached Massilia (Marseille) where they were rescued."

Again, it concerned a landing in Marseille, again in a ship without sail or oars. "From Marseille, Joseph and his company went to England, and after preaching the gospel there, they died."

Joseph's transit to England was recorded here. "These were the names of the castaways—Joseph of Arimathea, Mary the wife of Cleopas, Martha, Lazarus, Eutropius, (Mary) Salome,

16. de Voragine, *La Légende dorée*, XCV.1, p. 338–47; van der Meer, *The Black Madonna from Primal to Final Times*, 266n22.

Cleon, Saturninus, Mary Magdalene, Marcella (handmaid of the Bethany sisters), Maximinus, Martialis, Trophimus and Sidonius (elsewhere called Restitutus, the man born blind)."[17]

Note that Mary the mother of Jesus was not mentioned; she would have stayed behind. Other accounts mentioned other names such as James and Philip and still others mentioned Mary the mother among the passengers.[18]

It is important to determine the following. According to the reports of Rabanus, Jacobus de Voragine, and Baronius, ecclesiastical tradition transmitted that Joseph of Arimathea, Mary Magdalene, and the two sisters of Mary the mother, namely Mary Salome and Mary Cleopas—essentially Yeshua's immediate family—landed in Marseille.

Long before Baronius reported in the sixteenth century that Joseph was going to England, in the third and fourth centuries Eusebius, Bishop of Caesarea (260–340) and Hilary of Poitiers (300–367) wrote that immediately after the crucifixion some apostles went to Glastonbury. So it is confirmed by Cardinal Baronius that from France Joseph of Arimathea traveled on to England, therefore making that trip a historically and ecclesiastically recognized fact.

The Legends of Provence

In the nineteenth century, the Provençal poet Frédéric Mistral (1839–1914) collected a large number of Provençal legends and incorporated them into his long poem Mirèio (Mireille). It was published in 1859 and consisted of twelve songs. He wrote:

> After the first persecution, in which Saint James was slaughtered by the sword, his followers were put in a boat without

---

17. Baronius, *Annales Ecclesiasticae*, vol I, p. 327; Jowett, *The Drama of the Lost Disciples*, 33, 63, 70, 164; Prophet, *Mary Magdalene and the Divine Feminine*, 237.
18. Jowett, *The Drama of the Lost Disciples*, 63.

oars or sails off the coast of Palestine, somewhere near Mount Carmel, and left to fend for themselves and put away. In the boat were: Mary, wife of Cleopas (sister of Mary); Salome, often called Maria Salome (sister of Mary); Mary Magdalene; Martha and with the latter two their maid Marcella. They were accompanied by the following men: Lazarus; Joseph of Arimathea; Trophimus; Maximinus; Cleon; Eutropius; Sidonius (or Restitutus or the man born blind); Martialis and Saturninus. And as the boat drifted away, Sarah, the handmaid of (Mary) Salome and Mary Cleopas, threw herself into the sea to join her mistresses and with the help of Mary Salome she was brought into the boat.[19]

Here again, the story of the boat without oars or sails was confirmed. Mistral then gave an overview of places where the passengers of the floating boat ended up after landing in France. He also confirmed that Joseph of Arimathea traveled on to England.

After floating for several days, the boat drifted to the coast of Provence and following the Rhône, they arrived in Arles, which was converted to Christianity by the blessing of God and the preaching of Trophimus. Martha and Marcella went to Tarascon and Avignon. Martialis to Limoges. Saturninus to Toulouse. Eutropius to Orange. Lazarus to Marseille. Maximinus and Sidonius to Aix and St. Maximin. Mary Magdalene to St. Baume. Joseph (of Arimathea) is said to have crossed the sea to England. Maria Salomé, Maria Cleopas and Sarah, their maidservant, stopped at the seacoast of the Camargue and died there. And the small town of Les-Saintes-Maries-de-la-Mer, with the church where their relics are enshrined, keeps their memory alive.[20]

---

19. Taylor, *The Coming of the Saints*, 126–127; Jowett, *The Drama of the Lost Disciples*, 70.
20. Taylor, *The Coming of the Saints*, 126–127.

There are several traditions surrounding Mary, the mother of Yeshua. In Western Christianity she came along with Joseph of Arimathea and Mary Magdalene; this corresponded to the wish of Yeshua in *The Gospel of the Beloved Companion* where he entrusted the care of his mother to his wife.[21] It seemed logical that Mary the mother accompanied her brother Joseph and daughter-in-law Mary Magdalene to the south of France. However, in the Eastern Orthodox tradition, Mary the mother went with John to Ephesus and died there.[22] There is also a third tradition from Kashmir in India that she was buried in Murree.[23] According to channeled sources, Mary Anna was born in Ephesus and wanted to die there (chapter 5).[24] She may have traveled to Ephesus from the south of France. In that case, the Western and the Eastern traditions would both be correct.

How did the people of the Way disperse when they arrived in Marseille in France quite soon after the crucifixion? Based on the legends that Mistral compiled, it seemed as if every member of the Way went out on a lucky trip. But was it random distribution? Appearances are deceiving. There was indeed a system in the distribution. It was only very recently that I started to see the common thread. The line that connected them all was: the tin route.

A FRENCH INTERLUDE:
THE FIRST PEOPLE OF THE WAY
FOLLOWED THE TIN ROUTE

Joseph of Arimathea developed a network in which the tin from the mines in southern England was transported by ship

21. *GBC* [39:3], 73; *MMU*, 435–37.
22. Ralls, *Mary Magdalene*, chap. 5, "The Eastern Tradition."
23. Ahmad, *Jesus in Heaven on Earth*, 387 with photos of the tomb of Mary before and after the restoration on pages 371–75.
24. Heartsong and Clemett, *Anna, the Voice of the Magdalenes*, 368, 373.

to Morlaix in Brittany in France (see fig. 4.1). From there, with the help of packhorses, it traveled across France to the port of Marseille to be shipped to the Roman ports around the Mediterranean. Along the entire route through France, which ran through the valleys along various rivers, there were stopping places where the tin caravan spent the night.

After arriving in the south of France, many members of the extended family of Yeshua and Miryam spread across France along this tin route set out by Joseph of Arimathea.[25] So there was a system in the way France and England were won over by the people of the Way to the new vision of Yeshua and Miryam. In 1907, John Taylor gave a map on which the route of Joseph of Arimathea from Marseille through France to England was marked (fig. 4.1). It went from Marseille via Figeac and Rocamadour to Morlaix in Brittany. It was no coincidence that legends about Joseph of Arimathea were handed down in these places. In Morlaix, an estuary cuts deep into the land; at the end of this channel is the port of Morlaix. There the River Morlaix flows into that sea channel. That place can therefore be reached by inland vessel and seagoing vessel. From the seaport of Morlaix it went by ship to Ictis in Britannia (see fig. 4.1), where Joseph went to the land given to him on the island of Avalon, with the later village of Glastonbury.

The tin was brought across France by land to Marseille. The land route may have been traveled on foot in Roman times by people accompanying the packhorses. This was not insurmountable or impossible, given the time it took for a Roman legionary to get from Rome to Lutetia, or later Paris over a distance of 1,682 kilometers (1,045 miles). In the summer it took a Roman legionary (on foot at a marching pace) twenty-nine days

---

25. Taylor, *The Coming of the Saints*, 180, 266: "This well-known journey of the tin merchants presents no difficulty from the mouth of the Rhone to Cornwall . . . . The recognition of this route is almost certainly the route of the early missionaries."

to get from Rome to Paris. So that's a month on a tough route over the passes of the Alps. John Taylor provided the following possible timetable for the travelers wishing to travel from Marseille to Morlaix. From Marseille upstream on the Rhône to Arles, the journey through Gaul took thirty days. The sea journey across the Channel took four days. In Cornwall headed inland on a trade route to the Cornish tin mines and the lead mines in the Mendip Hills the journey took all together five to six weeks.[26]

Once I understood that the tin route was the common thread in the dispersal of the people of the Way, I took a map of France showing the river arteries. I started following the rivers from the south to the northwest. And with the reference points that Taylor provided on his 1907 map, namely Marseille, Figéac, Rocamadour, and Morlaix, I set out the other stops on the tin route (see fig. 4.1). Roads run through the river valleys and settlements were located along those roads. And it is precisely there that you will find the first evidence of the presence of the people of the Way. There, after centuries of oral survival, the historical facts developed into folktales and legends. Imagine a follower of the Way who wanted to travel from the south of France via western France to the coast of northwestern France and from there embark for England. In the far south, the Romans built good roads in record time, including Via Domitia, Via Aurelia, Via Agrippa, Via Julia Augusta, and Via Aquitania.

From Massalia or Marseille, the pilgrims walked via a well-passable Roman road to Aquae Sextiae or Aix. From there they follow the Roman road to Nemausus or Nîmes. Many rivers in southwest France flow to the east and north and often flow into the Garonne near Bordeaux. The pilgrims can therefore travel by ship over various rivers. Following the course of the River

---

26. Taylor, *The Coming of the Saints*, 179–80.

Figure 14.1. Map of Roman Gaul showing Roman roads to major cities.
Roads run through the river valleys and settlements were located
along those roads.

Aveyron you reach Segodunum or Rodez. (The Aveyron is not on
this map but flows parallel between the Lot and the Tarn into the
Garonne.) Next follow the River Lot and you reach the town of
Figeac and from there Rocamadour, also on the Lot, and further
northwest again is Divona or Cahors, also on the Lot. From there
it goes via the Isle, a tributary of the River Dordogne to Vesunna
or Périgieux. Via the River Vienne it goes to Augustoritum or
Limoges. Follow the River Vienne north and reach Limonum
or Poitiers. The pilgrims continued along the Loire and reached
Angers and further along the Loire Condevicnum or Nantes.

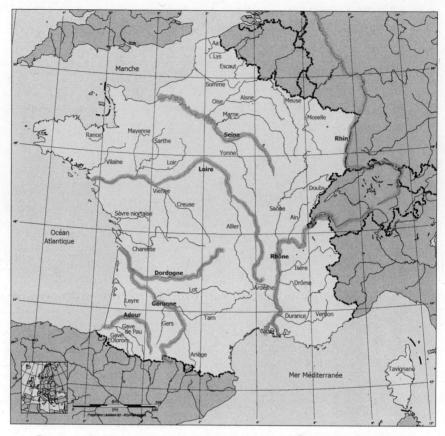

Figure 14.2. Map of rivers in France with the most important cities on those rivers. The River Aveyron is not on this map. It flows with tributaries between the Lot and the Tarn. Author: Lvcvlvs.

These pilgrims reached a final stop along the Loire: the city of Condate or Rennes. Finally, they followed the River Morlaix to the port city of Morlaix on the Bay of Morlaix.

But what proves that the French port of Morlaix had also been used as a port under the Romans? Olivia Hulot discovered that answer in 2018 with her subaquatic archaeology team. That year at a press conference Olivia reported on the finds she found at the bottom of the sea off the coast of Morlaix. She and her team found a Roman tin ship from the third to fourth century in the bay of Morlaix at a depth of forty-five meters. The

Roman freighter was found broken in two off the coast near the rocks of the Île de Batz. Five hundred ingots (blocks in which the tin was cast) were unearthed. The tin was probably mined in Cornwall and Devon. Numerous photos can be found on the internet showing the ingots lying on the bottom of the sea (see fig. 13.1 for an example of copper ingot from Cyprus).

From the Bay of Morlaix, the pilgrims crossed the English Channel. There are good maps showing various ports in southwest England that were under Roman rule after 47 CE. It is theorized that one of the reasons for the Romans to conquer England was to give them easy access to the tin and lead mines. From Morlaix, the pilgrims traveled via the Isles of Scilly to St. Michaels Mount and other ports in southwest England (see fig. 4.1).

Legends circulated in France around the people who fled from Palestine in a boat, and some sources mentioned three boats. Together with the surviving historical sources, these legends form the basis for the collections of the lives of the saints, which have recently been reissued.[27] Around 1900 John Taylor meticulously researched the many local and regional legends in France and England. If you put all this information together, the following picture emerges about the places in France and England where the first people of the Way stopped for some period of time or settled permanently. I went into this in my book *The Black Madonna from Primal to Final Times.*

Tracing the Women's Journey

- Sarah and the two Marys (Mary, wife of Cleopas, and Mary Salome—aka Mary the mother's sisters) went to Les-Saintes-Maries-de-la-Mer in the Camargue.[28]

---

27. Etienne Michel Faillon, ed., *Monuments Inédits Sur L'apostolat De Sainte Marie-Madeleine En Provence Et Sur Les Autres Apôtres De Cette Contrée, Saint Lazare, Saint Maximin, Sainte Marthe Et Les Saintes Maries Jacobé Et Salomé*, Vol II, Paris: Chez l'editeur, 1848.

28. Taylor, *The Coming of the Saints*, 246–49.

- Mary Magdalene and her brother Lazarus can be traced to Marseille, Aix, Sainte-Baume,[29] and in England.[30]
- Martha and her maid Marcella went to Arles and later to Tarascon and Avignon.[31]
- According to some English reports, Mary the mother was buried together with Joseph of Arimathea in Avalon in England.[32] According to the channeled sources and Eastern tradition, she traveled from the south of France to Ephesus and was buried there.

Tracing the Men's Journey

- Joseph of Arimathea left traces in Limoges, where he passed in transit with the tin caravan.[33] And the same went for Rocamadour; he passed through there with Zacchaeus. Furthermore, Joseph left traces in legends in Morlaix[34] and in England he can be traced to Cornwall and Glastonbury.[35]
- Lazarus spent a long time in Marseille.[36]
- Trophimus traveled to Arles and the entire Narbonne district with Arles as its center. He was the first bishop of Arles.[37]

---

29. Taylor, *The Coming of the Saints*, 240–46.

30. van Dijk, *Maria Magdalena, de Lady van Glastonbury en Iona*, 148, 149 states that Mary Magdalene visited England.

31. Taylor, *The Coming of the Saints*, 249–52.

32. Jowett, *The Drama of the Lost Disciples*, chapter 13, "Did the Virgin Mary Live and Die in Britain?" 132–48, the author answers this question in the affirmative.

33. Taylor, *The Coming of the Saints*, 174.

34. Taylor, *The Coming of the Saints*, 174, 224.

35. Taylor, *The Coming of the Saints*, chap. 8 "St. Joseph and Glastonbury," 173–216 and 224.

36. He would have lived there about fifty years and been the leader of the Christian congregations there. See Taylor, *The Coming of the Saints*, 238–240; Jowett, *The Drama of the Lost Disciples*, 163: the relics of Lazarus have been found there, where he was bishop for seven years.

37. Taylor, *The Coming of the Saints*, chap. 7 "St. Trophimus and Arles," 150–72.

- Maximinus and Sidonius went to Aix and Saint-Maximin.[38]
- Cleon traveled all over Gaul.[39]
- Eutropius can be found in Orange.[40]
- Sidonius (his other name is Restitutus or the Blind Man)[41] traveled with Cleon and Eutropius throughout Gaul.
- Martialis moved to Limoges[42] and Avalon with his parents, Marcellus and Elizabeth.
- Saturninus went to Toulouse.[43]
- Maternus went to Trier.[44]
- Zacchaeus went to Rocamadour.[45] He was permanently linked to Rocamadour in the legends surrounding the Black Madonna of Rocamadour.

Figure 14.3. The Black Madonna of Rocamadour. Photo by Jaap Craamer.

38. Taylor, *The Coming of the Saints*, 240–44.
39. Taylor, *The Coming of the Saints*, 126.
40. Taylor, *The Coming of the Saints*, 105, 106, 107, 112, 126.
41. Taylor, *The Coming of the Saints*, 252–55.
42. Taylor, *The Coming of the Saints*, 261.
43. Taylor, *The Coming of the Saints*, 107, 126.
44. Taylor, *The Coming of the Saints*, 266–71.
45. Legends surrounding Zacchaeus in Rocamadour, the Jericho tax collector, are mentioned in Lk. 19:1–10; Taylor, *The Coming of the Saints*, 224, 261–65.

Remember that Gaul only became a Roman province in 27 BCE; the resistance to the Roman occupation of Gaul lasted until the year 70 CE. Before that, several diasporas from Palestine went to Gaul. Many Jews lived in France, and England only became a Roman province in the year 43. Roman domination lasted there until 410. It must have been easy for refugees to go into hiding in these areas, which were not (or hardly) Romanized during the lives of Yeshua and Miryam.

15
THE TRACES
MARY MAGDALENE
LEFT IN FRANCE

Mary Magdalene taught and traveled in France, and many legends hold her in high esteem. She was regarded as the most important evangelist of France. She was even considered to be the patroness of France.

In the Footsteps of Mary Magdalene

Mary visited many cities.[1] Her trail began in Marseille, then briefly went to Aix, Arles, and Tarascon, where her sister Martha settled. There are places in the landscape around Rennes-le-Château with the legends woven around it, which indicates that she was active there. From there it goes to Vézelay and onto its penultimate stop in a high cave near Sainte-Baume. Finally, she was buried at the foot of the mountain massif. Her black skull was kept in the crypt in the current basilica. Many local legends were the distant and sometimes distorted echoes of long-gone historical events. Let's walk the mapped route step by step. In my book *The Black Madonna from Primeval to Final Times* I provided the images that belong to this journey; I refer to the pictures in this book.

---

1. van der Meer, *The Black Madonna from Primal to Final Times*, ch 7.6 "Mary Magdalene at work in the South of France" and following paragraphs, 278–91.

Marseille

The Life of Mary Magdalene by Jacobus de Voragine stated Mary Magdalene and her brother Lazarus were not allowed into Marseille. Lazarus and Mary settled on the south side of the old city, on a hill with a cave and a well. This place outside the city was the necropolis of Marseille, the place where the dead were buried. Between the graves and tombs they cared for the sick and helped the poor, and Mary taught and spoke in public. "And all admired her both for her eloquence and for her beauty."[2] Initially, Lazarus and Mary were not allowed to enter the city of Marseille, but the local rulers dreamt about Mary for three consecutive nights, resulting in permission for both to enter.[3] It is said that the people of Marseille were deeply impressed by their actions.

In the second century, a chapel called Notre-Dame-de-la-Confession was built over the cave where Mary settled with her brother. A Black Madonna of the same name later chose her domicile here. The chapel was later developed into the church of the Abbey of Saint-Victor. The current lower church is formed by the grotto and the chapel. The church is, as it were, built over and around it. The monks of the Abbey of Saint-Victor later claimed that they had important relics of Mary in their possession, because Mary would have lived in the cave, which was converted into the crypt of the church in the third century. There would also be the altar where Mary preached after her arrival.

In the crypt, as mentioned, the Black Madonna called the Notre-Dame-de-la-Confession is venerated. On February 2, the feast of Candlemas and the beginning of spring, large groups of pilgrims descend to the crypt while it is still dark. They light green candles and carry the Black Madonna statue, wrapped in a green processional dress, up to the light. While still partly in the dark, they make a tour through

---

2. de Voragine, *La Légende dorée*, Vie de Sainte Marie-Madeleine, pècheresse, n. XCV, 338–47, c. I, 340.
3. de Voragine, *La Légende dorée*, c. II, 340.

Figure 15.1. Mary Magdalene lands in Marseille and addresses
the inhabitants there. *The Speculum Historiale*
by Jean de Beauvais from 1463.

the awakening city (fig. 15.2). Finally, the statue is placed in the dark
crypt again for a full year. The inhabitants of Marseille feast on the
navettes, cakes in an elongated barque shape, which symbolize the
ship in which Mary Magdalene and her company arrived in Marseille.
These look like elongated minicanoes and resemble the barque of Isis
(fig. 15.3). Marseille has an old tradition around Isis, Mary Magdalene,
and the Black Madonna.

Figure 15.2. The Black Madonna of Confession or Notre Dame-de-la-Confession is normally located in the crypt of the church of the Abbey of Saint-Victor in Marseille. On February 2, she is brought up from the crypt early in the morning and carried around the city. Photo by Jaap Craamer.

Figure 15.3. The special pastries in the shape of the barque of Isis. Photo by Jaap Craamer.

Aix-en-Provence

From Marseille, Mary Magdalene was said to have traveled to nearby Aix. The disciples Maximinus and Sidonius had traveled here earlier, and she would also teach there. Aix is known for several hot and cold springs; the name Aix comes from the contracted Latin name of

Aquae Sextiae. The Basilica of Sainte-Marie-Madeleine and a cathedral were later built in the city. Then Mary continued her journey to Arles and to Tarascon where Martha had settled; numerous legends were woven around this place too. Then she went to Rennes-le-Château.

Rennes-le-Château

The city is located on a mountain and is intertwined with Mary Magdalene. The area is known for its hot springs, the salty River Sals, rich mineral resources, and the abundance of water. Folktales describe how Mary healed the sick here, as well as teaching and baptizing. Three special places are mentioned in the surrounding landscape: the Source of the Mother Goddess at Campagne-les-Bains; the Source of the Lovers or the Fontaine-des-Amours; and the holy water font or *le benitier*. It is at this last place where Mary was said to have baptized people. These days we also know the cave of Mary Magdalene, a birth cave with a vulva-shaped opening right next to it.

In the town of Rennes-le-Château, the priest Bérenger Saunière, who was appointed parish priest from 1885 to 1917, carried out numerous reconstructions around the church on the citadel. He had the Mary Magdalene tower (the Tour Magdala) built for his library. The parish priest had also completed extensive research in the area. Saunière spent a fortune at the end of the nineteenth century on the renovation of the church, parsonage, and surrounding gardens as well as the construction of the tower, which rises high over the deep plains below. Some people suspect that he brought things to light that earned him the large amount of money to pay for all this.

On this high place stood a castle with a house chapel; the chapel was dedicated to Mary Magdalene in 1095. Later the chapel was converted into a parish church, and it was also dedicated to Mary Magdalene. Mary Magdalene is often portrayed in the existing church.

The former Mary Magdalene chapel probably rested on older foundations of a Visigoth fortress. The Visigoths ruled France in the fifth and sixth centuries. They plundered Rome and may have captured treasures that they hid in Rennes-le-Château. They were driven out by the Franks, who were still nominally headed by the royal house of the Merovingians, a house that had always had a special relationship with Mary Magdalene: the Merovingian king Clovis visited the tomb of Mary Magdalene in 480. There was still a hint of mystery surrounding the alleged treasure and Mary Magdalene in Rennes-le-Château (chapter 17, also about the Visigoths).

Sainte-Baume

Mary Magdalene would not have spent her last years in Rennes-le-Château but in Sainte-Baume—or at least that is what local lore in Baume says. The Massif de la Sainte-Baume is a twelve-kilometer-long ridge, with an average altitude of one thousand meters and located twenty kilometers northeast of Marseille. If you follow the river L'Huveaune inland to the north, you will get there. The springs of the river rise near the massif. There is a lot of rain here, and all that water flows through the sources of the river and the river itself along the most enchanting places and out into the sea. The road from Marseille to Baume runs along the river through this unspoiled landscape. Mary must have gone this way before retreating to a cave in this massif. The legend says that every day the angels carried her here to heaven where she was fed.[4]

The traditional Catholic version says that after a short stay in Marseille and Aix, Mary Magdalene spent thirty-three years here as a hermit in solitude.[5] But there are too many memories of her in the rest of the south of France to justify such a long stay in complete isolation. In all likelihood she had been confused with the hermit Mary, an Egyptian of the late fourth to early fifth centuries.[6] This Egyptian Mary had led a dissolute life as a hermit and was often depicted with long, loose hair, but she became identified with Mary Magdalene. The

---

4. de Voragine, *La Légende dorée*, XCV op 338–47, c. III, 343.
5. de Voragine, *La Légende dorée*, XCV op 338–47, c. III, 343 for the passage from The Life of Saint Mary Magdalene where it says she lives the life of a recluse for thirty years; see de Voragine, *La Légende dorée*, LVI, 212–215 for The Life of Saint Mary the Egyptian.
6. Rabanus Maurus denies that Mary Magdalene would have been a hermit. (See previous chapter.)

traditional image of Mary as a hermit and penitent martyr distracted attention from her actual performance, in which she was recognized as the patroness of France. When Mary had not withdrawn into the cave for thirty-three years, she had been active in France and England from her arrival after the crucifixion until the year of her death, probably in 63 CE. Mary Magdalene probably only retreated at the end of her life to a cave high in an inaccessible mountain massif just above Marseille, or she did this to build in periods of rest in her busy life.

Saint-Maximin-la-Sainte-Baume

When Mary felt that the end was approaching, she was said to have gone to Saint-Maximin-la-Sainte-Baume; upon her death she was laid in a tomb there by Saint Maximin. The date of her death would have been in 63. This was recorded in the Vatican *Desposyni* genealogy presented to Pope Sylvester I in 318; this family tree of the family of Jesus is in the Vatican archives. When she died, Mary Magdalene may have been sixty years old.[7] Her remains were guarded by the monks of the order of the Cassianites. The founder of this order was Cassianus (360–435). He certainly took numerous documents from Bethlehem in Palestine in the early fifth century and settled in the south of France. He initially resided in the Abbey of Saint-Victor. Later he built a chapel in Sainte-Baume on the place where Mary Magdalene was buried and later on, a basilica with an abbey was built there. Her cemetery in Saint-Maximin was guarded by monks of the Cassianite order until the eighth century.

Because the Saracens from Spain threatened to occupy the place, her remains were moved to the sarcophagus of Saint Sidonia. In an accompanying document, the monks stated that they were doing this to protect her remains from grave robbery and destruction. According to others, her remains were moved to Vézelay, where a beautiful basilica dedicated to Mary Magdalene was built in 1194. Vézelay subsequently became a famous place of pilgrimage. But in 1279 you can find again in the crypt of Saint-Maximin a tomb with the remains of Mary and her

7. Gardner, *The Magdalene Legacy*, chap. 2.

Figure 15.4. Gold bust from the crypt of
Saint-Maximin-de-la-Sainte-Baume with the jet-black skull
of Mary Magdalene. She wears a veil of gold and precious
stones and has her "face" covered with glass.
Photo by Jaap Craamer.

now blackened skull. Vézelay may have been able to obtain some relics, but the tomb of Mary Magdalene is in Sainte-Baume.

You can still admire her skull in the crypt of Saint-Maximin-de-la-Sainte-Baume. The skull, which has turned pitch black with age, is contained in a golden bust, wears a crown of gold with precious stones, and has the "face" covered with glass (fig. 15.4). French monarchs regarded her as "a daughter of France." Her tomb was visited by many crowned heads, including, as already mentioned, the Merovingian King Clovis in 480 as one of the first.

In the crypt in the Basilica of St. Maximin there is a very old inscription on a stone from about 375. Inscribed is Mary Magdalene

MARIAVIRCO
MINESTERDE
TEMPVLOCEROSALE.

Figure 15.5. Sketch of
inscription on stone.

in the *orante* position: a praying body position with outstretched arms. Above it is the following text in Latin which reads in translation: "The Virgin Mary, Officiating Priestess in the Temple at Jerusalem."[8]

Mary was called priestess of the temple in Jerusalem. What a miracle to find this stone next to her skull in the crypt. The figure shows a sketch of the full inscription (fig. 15.5). The traditional interpretation holds that it was Mary the mother; but I think this stone next to her tomb proves that it is Mary Magdalene, the patroness of France.

Traces of Mary in the French Landscape

Some legends are entrenched in numerous landscape features. There are Mary Magdalene mountains, caves, and springs. It was mentioned at

---

8. Edmond Le Blant, *Les Sarcophages Chrétiens de la Gaule*, Paris: Imprimerie Nationale, 1886, 57 with Latin version: *Maria Virgo Minester de Tempulo Hierosale*.

Rennes-le-Château. Together with the Black Madonna, Mary left traces in the landscape. The landscape and the energetic values of places of power in that landscape are studied by sacred landscape sciences and modern geomancy.

Mary's Connection with the Black Madonna

Often in France the paths of Mary Magdalene and the Black Madonna cross. An example is the grotto in Marseille, where she is said to have lived and preached for a while after her arrival in France. As discussed earlier, a chapel with a crypt was built over that cave in the second century and later, a large church, the St. Victor, was built over it. In the crypt of that church, the former cave of Mary Magdalene, you will find a Black Madonna. There are rituals around this Black Madonna on traditional Marian festivals. In Marseille on February 2 during the spring festival of Candlemas, people commemorate that Mary Magdalene once landed in Marseille by baking and eating navettes (biscuits shaped like Isis's barque).

There is a connection between the Black Madonna worship, Isis worship, and Mary Magdalene. Numerous Isis statues have later been transformed into Black Madonna statues. In addition, France has numerous other dark goddesses, such as the black Ana of the Celts; the black Artemis; the black Aphrodite; and the black Athena of the Greeks. Their dark aspect was worshipped in caves and crypts. The goddess in her dark aspect gave birth to new life, new light. It was the regenerating side of goddesses like Ana, Isis, and others. That regenerating power of the ancient mother goddess can be found in the Black Madonna worship. It is related to the performance of Mary Magdalene; she brought to France a hidden and later forbidden Christian mystery teaching, a teaching that is allowed to come to light again in this time. In the eyes of the Dutch author Jaap Rameijer, this makes France the second Holy Land, the land of Mary Magdalene and the Black Madonna.[9]

---

9. Rameijer, *Maria Magdalena in Frankrijk*, 68.

France has a Living Mary Magdalene Veneration

France has an existing and therefore living Mary veneration. Take the tradition of the town of Soursac. On Mary's name day on July 22, people from this town go on a pilgrimage to the higher chapel above the village of Nauzenac in the Corrèze department. A large statue of Mary Magdalene, carved from dark wood, is carried along. At that chapel in the fifteenth century, a white lady appeared to two young shepherds who were looking for lost lambs. She introduced herself to them as Mary Magdalene. That was why this luminous lady with the dark face is called The Lady of the Light or La Dame de la Lumière in France. The site is near the ravine or gorge carved long ago by the River Dordogne. There is a living veneration of Mary Magdalene there, deeply rooted in the collective consciousness of the peasant population.

Figure 15.6. Recent procession in which the statue of Mary Magdalene was carried around. Photo by Jaap Craamer.

Mary Magdalene is Very Present in the French Art Tradition

Nowhere in the world are there so many churches dedicated to Mary Magdalene as in France; nowhere in the world have so many statues, panels, and paintings about her been found. How is she portrayed?

- In France you come across a heavily pregnant Mary Magdalene as she stands on the high altar of the Sainte-Marie-Madeleine church in Paris. Numerous other examples can be found in France.
- Sometimes, heavily pregnant, she holds a small child by the hand or on her arm.
- You can recognize her in pictures by her blood-red robe—sometimes with a green outer garment—whether or not she holds a child on her arm or by the hand. She is also depicted like this on Eastern icons. Starting in 1649, the Inquisition determined a new color code in the Western Church: Mary the mother must be depicted in white and light blue, or black, and Mary Magdalene must be depicted in red, the color of lust, and green, the color of nature. That made Mary the bride and wife of Jesus easy to recognize.
- Finally, there is the tradition in which she is depicted as Jesus's companion during the Last Supper. There is a well-known sculpture from the twelfth-century Saint-Volusien church in the town of Foix in the Ariège department in the Cathar country in France. The tradition in which the beloved disciple from the Gospel of John is depicted next to Jesus as a beardless female figure with long red hair is very present in the art tradition in France. This tradition is later found in *The Last Supper* by Leonardo da Vinci (1495–1498). This must be based on apocryphal texts that circulated in France and Italy about Mary Magdalene. It was confirmed by *The Gospel of the Beloved Companion of Jesus*, where Mary Magdalene (writing about herself as the Beloved Companion) had a place of honor at Jesus's breast. This makes it easier to understand why Mary Magdalene

Figure 15.7. *The Last Supper with Jesus and Mary Magdalene.*
Sculpture from the twelfth-century Saint-Volusien church in the
town of Foix in the Ariège department of France.

is so prominent in the art tradition in Western Europe and
France, Italy, and England in particular.

From Fantasy and Fiction to Historical Fact:
The Regression Reports

The fact that Jesus and Mary Magdalene would have been married has
recently been discussed in bestsellers such as *The Da Vinci Code*. Prentis
asked Alariel the following question: "We have noticed that much of
the controversial material about Mary Magdalene is presented in a form
of fiction. Would you mind commenting on that?"

Alariel replied, "When an idea is particularly challenging, your culture often needs to explore it first on a fictional level before moving on
to accept it as fact. At the level of fiction—and even more so on the
level of fantasy—your consciousness is able to accept these ideas without disturbing your belief system. This will allow you to ponder these

ideas and become accustomed to them. When you come to the conclusion that they are not as dangerous as you thought, only then will they move into the realm of fact and be widely accepted throughout your culture. In this way, your artists help you to broaden your belief system and move your culture on. A general principle emerges from this: 'What is seen as today's fantasy may be accepted as tomorrow's fact."[10]

10. Wilson and Prentis, *Power of the Magdalene*, 126–28.

16
MARY MAGDALENE
IN BRITAIN

Mary Magdalene was said to have traveled from France to England, where the priestesses of Avalon reside to this day.

Mary Magdalene and the Priestesses of Avalon: The Regression Records

Prentis asked Alariel to tell her something about the priestesses of Avalon. Alariel answered,

> The priestesses actually predate the Druidic impulse, as they are essentially an offshoot of the Lemurian Mother-Energy and Mother-Wisdom. They were able to receive this tradition because the Lemurian wisdom was kept alive by the Melchizedek teachers and then taught by the Kaloo.
>
> ... Mary Magdalene, as a high initiate of the Isis mystery school, came to Avalon and opened up higher levels of the mysteries and a deeper interpretation of the Mother-Energy as it had developed within the Isis tradition in Egypt. This reconnection at a higher level gave the mystery school, guided by the priestesses of Avalon, a completely new infusion of life, and the subsequent teaching and initiation of many by the Order made bright the last stage of the Druidic impulse in Britain. Through Mary they were able to

reconnect and re-empower the Order, so the priestesses became recognized as a fount of empowered wisdom capable of balancing the power and wisdom of the Druids. The balancing of masculine and feminine energies achieved in this way paralleled the balance achieved so many centuries earlier on Atlantis and was widely recognized as being a remarkable achievement by the wise ones throughout the Celtic world at this time.

Mary Magdalena grasped the fundamentals of the Isis tradition so well . . . that she was able to interpret the existing rituals and teachings of the priestesses of Avalon from a higher and more esoteric standpoint. Up to the point of Mary's arrival in Avalon, the priestesses had some keys to the mysteries, but much of their own teaching was still veiled and hidden from them. Mary brought the inner essence of these teachings out into the light and enabled the priestesses to understand their own tradition in a much more profound way. For someone to come and make sense of rituals which you have been practicing for generations, so the deepest resonances and symbolism within these rituals suddenly become clear to you, was a very powerful and inspiring thing. And this was why the priestesses of Avalon were able to accept Mary so quickly. She revealed connections and resonances and a profound symbology that had always been there in the rituals but were not previously recognized. It is not too much to say that Mary's interpretation came as a revelation to the priestesses, who thought they knew and understood their own rituals perfectly well. In this way she became a legend during her lifetime. She brought the fire of wisdom down from the most subtle levels of consciousness and used it to help and heal. Mary had the ability, through her words, her energy and her unique presence, to communicate the nature and the essence of the All.

If you consider Mary's work outside Israel, her reinvigoration of the Order of the priestesses of Avalon could be considered as a major achievement. There are some parallels here with the work of Yeshua. As he was reforming the guru system in India, Mary Magdalene was at about the same time doing very similar work in Britain through the priestesses of Avalon. They were both reforming and reinvigorating an existing tradition through the application of wisdom at a deeper and more profound level. Although Mary returned to continue her work in what is now France, the time she spent teaching the teachers at Avalon had a profound and beneficent effect on the whole Druidic civilization. While the Druids revered Yesu, the Coming Savior, that Savior was not walking among them after the crucifixion, whereas Mary, a high initiate with a unique link with Yeshua and unique insight into his teachings, was available for a time at Avalon to teach, inspire and guide. The teachers she taught inspired a whole generation of initiates in Britain and from many other parts of Europe. In many ways Mary's work provided the last and most elevated high peak of the long and illustrious Druid tradition.[1]

The Three Marys of Glastonbury: Traces in Geography and Iconography

In Avalon, Mary the mother, Mary Magdalene the bride, and the Black Madonna left their mark. The current priestesses of Avalon carry on the Marian tradition at Glastonbury. Kathy Jones, a modern priestess of Avalon, pointed out that Mary appeared in Avalon in three forms.

- In Mary the mother to whom the wattle church or Old Church is dedicated.
- In Mary Magdalene after whom several places, plus a street in and around Glastonbury, are named.
- In the Black Madonna once worshipped at Glastonbury in the form of a very old, black, wooden statue. The triple goddess from matriarchal-megalithic times in her tricolor of white, red, and black returns in the three Marys.[2] There is no battle over which of the three is the right one here—the three Marys may all be.

---

1. Wilson and Prentis, *Power of the Magdalene*, 116, 117–118.
2. Jones, *In the Nature of Avalon*, 134.

They exist side by side in threefold unity. The triple goddess, in whom the three are one, expressed herself in the landscape as a giant lady, worthy of her megalithic past.

The Landscape Goddess

The megalithic people who inhabited Ireland and England prior to the arrival of the Celts had a great reverence for Mother Earth. They knew her in her threefold aspect of Life–Death–New Life: bestowing, taking, and giving life again. This world view of eternal renewal is symbolized by the three colors of white, red, and black. The white aspect of the Mother is experienced in the early spring life. Her red aspect is in the full and fertile summer when nature is in bloom and can be harvested. Her black aspect is experienced in the dying of nature in autumn and winter and with the return of light after the winter solstice. The new light is born again from the darkness. Thus, each annual cycle continues to spiral through time and space.

The matriarchal peoples were star oriented. They built structures of large stones or megaliths as observatories to study the cosmic cycles of the stars. They experienced the dark cosmos as the garment of the goddess; her dark robe was studded with twinkling stars. But they also experienced the goddess in the earthly landscape. In the landscape, they saw the maternal forms of the ever-giving goddess who constantly renewed life. This legacy is carried on by the modern priestesses of Avalon, who have their center in Glastonbury. They honor Mother Earth in the landscape goddess.

In her 2007 book about the landscape in Avalon, Kathy Jones described the island when viewed from the south as not only resembling a swan in flight (fig. 13.3), but also a pregnant lady lying on her back in the green meadows of Summerland[3] (fig. 16.1). Her head, shoulders and right arm sink into the earth at the folds of Stonedown, a low hill to the east of the Tor. The Tor itself forms her left breast; the nipple pro-

---

3. Jones, *In the Nature of Avalon*, 21, 146; Mann, *The Isle of Avalon*, 126.

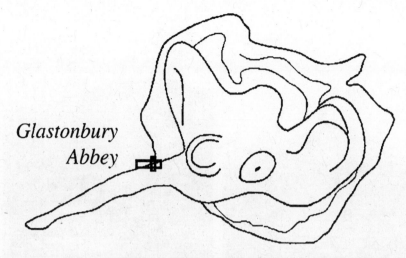

Glastonbury
Abbey

Figure 16.1. The reclining, megalithic landscape goddess with, at the vulva, the Glastonbury Abbey that once housed the wattle church.
From: Kathy Jones, *The Nature of Avalon*, 146.

trudes, so to speak, in the shape of St. Michael's tower, which is visible from afar. The right breast slips slightly as it does with a woman lying on her back. The round Chalice Hill forms her pregnant belly. Her left leg is extended into Wearyall Hill with her knee slightly bent. Her right leg is located at St. Edmunds and Windmill Hill. You see a gigantic landscape goddess from whose sides two springs flow: one spring gives white water and the other spring red, ferruginous water. From her pregnant womb she gives birth to life.[4] Glastonbury Abbey is located at the entrance to her vulva.[5] The area surrounding Glastonbury is sprinkled with megalithic structures.[6] The landscape goddess recalls those ancient megalithic times when the feminine was valued and balanced with the masculine.

Behind the spot where the little wattle church once stood on the site of the later abbey, the rounded Chalice Hill looms in the distance, at least in winter when the trees in sight bear no leaves (Figs. 16.2a–b).

---

4. Jones, *In the Nature of Avalon*, 21 with a clear sketch on p. 20; Jones, *The Goddess in Glastonbury*,; Mann, *The Isle of Avalon*, 126n20; Jones, *Priestess of Avalon*, 48, with an image of the landscape goddess on p. 52.
5. Jones, *In the Nature of Avalon*, 146.
6. Michell, *New Light on the Ancient Mystery of Glastonbury*, 27 and further.

Figs. 16.2a and 16.2b. The Tor. My photos from 2017.

Beyond that is the great hill called the Tor. It was a High Place of Worship of the Druids, a *gorsedd*. It was a hand-thrown earthen mound of greater size than the pyramids in Egypt, according to Jowett.[7] In such high and sacred places, the Druids had their astronomical observatories from which they studied the heavens. In ancient times they were known for their great astronomical knowledge. The Tor is an ancient, elongated, terraced hill of limestone, clay, and loam covered with grass.[8] The road along the seven different terraces leads to the top, where you have a magnificent view over the reclaimed land in the depths. In fact, you walk a labyrinth with seven turns, and it takes three to four hours if you want to make the full, oval circuit from bottom to top. Then, you also walk down from the labyrinth in three to four hours. The elongated Tor has the pattern of the Cretan labyrinth with the seven turns, a pattern that occurs universally[9] (fig. 16.3).

7. Jowett, *The Drama of the Lost Disciples*, 77.

8. van Dijk, *Maria Magdalena, de lady van Glastonbury and Iona*, 118.

9. Jones, *In the Nature of Avalon*, 160. She connects the circumambulation with the elements and with the seven chakras.

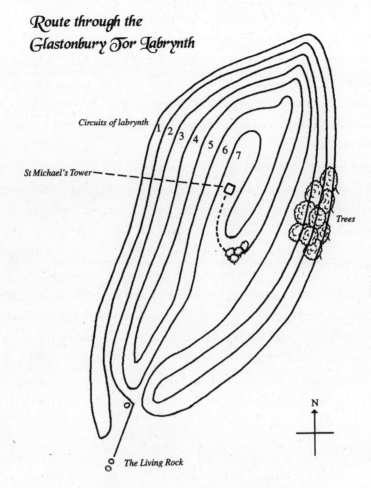

Route through the Glastonbury Tor Labrynth

Circuits of labrynth

1 2 3 4 5 6 7

St Michael's Tower

Trees

N

The Living Rock

Figure 16.3. The Tor as labyrinth.
From Kathy Jones, *The Nature of Avalon*, 161.

Now we'll follow the traces of the Three Marys.

The Black Madonna of Glastonbury

In the later, thirteenth-century version of William of Malmesbury's work, in which the monks of the abbey took the liberty of inserting all kinds of additions, you read that Joseph of Arimathea had a special dream in which the archangel Gabriel instructed him to build the first church in the British Isles, a wattle church, to be dedicated to the Virgin Mary. In addition, Joseph must carve an image of Mary and place it in the round wattle church to be consecrated to Mary the mother of Jesus. This statue became the legendary Black Madonna of Glastonbury. It stood in the Old Church until the Great Fire of 1184 (chapter 13). According to Kathy Jones, it was the only object that was saved and was placed in the new building when the church was rebuilt in 1186.

Immediately afterward, Glastonbury Abbey became an even more important place of pilgrimage. The most important holiday (which coincided with the most visits by pilgrims) is September 8, the commemoration and feast day of Mary's Nativity. The statue was lost forever when Glastonbury Abbey was destroyed by Cromwell's armies in 1539, yet the Black Madonna had not completely disappeared. There was a fourteenth-century seal from Glastonbury Abbey of Our Lady of Glastonbury. Figs. 16.4a and 16.4b show a sketch and a photo from the abbey's museum.[10] In the nineteenth century, a wooden statue called Our Lady of Glastonbury was made after the example of this seal. In the left hand she holds the child, and in the other she holds a staff with a blooming rose bush. She resides in the Catholic Church on Magdalene Street.

Mary the Mother of Yeshua in Avalon

The wattle church or Old Church is, according to legend, dedicated to Mary, the mother of Jesus.[11] In the following, Kathy Jones, the priestess of Avalon, is again our guide. In the body of the landscape goddess, there is a vulva. The territory that Joseph was given and where later the Old Church and later the abbey arise, lie in the vulva of the giant Lady, between her spread legs[12] (fig. 16.1).

After the fire, the first building to be rebuilt is the 'Lady Chapel,'

---

10. Jones, *In the Nature of Avalon*, 140, 153. Documents show that there was a chapel in the abbey where a Black Madonna statue stood.

11. Jowett, *The Drama of the Lost Disciples*, 75.

12. Jones, *In the Nature of Avalon*, 146.

Figure 16.4a. Photograph of a fourteenth-century seal from Glastonbury Abbey of a standing Virgin and Child. In one hand she holds a staff with a blooming rose bush. She is Mary of Glastonbury. My photo, taken in the Abbey Museum.

Figure 16.4b. A standing Mary holds her child in her left arm, and in her right hand she holds a staff that ends in a blooming rose bush. Photo from the Abbey Museum.

in exactly the same place as where the Old Church stood in the former Mary Chapel (fig. 16.5). In the crypt of the Lady Chapel there is a well (figs. 16.6a–b; fig. 13.6) with ground plan of the abbey. According to archaeologist Professor Philip Rahtz the masonry around the well was the oldest of the whole excavation and possibly pre-Christian.[13] The source itself certainly existed in pre-Christian times and, like all waters among the Celts, must have been dedicated to a lady. This spring was near where Joseph and his eleven companions built the round core structure with their twelve houses around it. The spring itself lays low and outside the wall of the crypt of the former Mary Chapel. From the source there was a direct view to the belly-shaped Chalice Hill with the much higher Tor beyond. Until a few years ago, people had free access to the source, and you could drink the water, but the source is now behind bars. With special permission, the gate swings open and you gain access to the source. According to Kathy Jones, the crypt with the spring is one of the most powerful energy places in Glastonbury.[14]

If you go out of the crypt via a staircase to the ground floor of the ruins of the Lady Chapel, you will find yourself between the high-standing walls of what is left of the Lady Chapel (fig. 16.5). Here—again according to Kathy Jones—a strong energy vortex can be experienced and observed between Heaven and Earth.[15] This high-quality energy is why the megalithic agricultural cultures, and their Druids honored this holy place with the source, followed by the first people of the Way.[16]

According to local tradition, Mary died and was buried in Glastonbury fifteen years after the foundation of the Old Church. She was said to be buried in the cemetery next to the Lady Chapel. In the weathered walls of the ruin you find a stone protrudes from the south

---

13. Rahtz, *Glastonbury*; Geoffrey Ashe, *The Discovery of King Arthur*, Garden City, NY: Anchor Books, 1985. The well is now named after Joseph of Arimathea, but according to Kathy Jones a better name would be "the Mary well."
14. Jones, *In the Nature of Avalon*, 147–48.
15. Jones, *In the Nature of Avalon*, 148.
16. Jones, *In the Nature of Avalon*, 148.

Figure 16.5. The Lady Chapel, my photo from 2017.

Figures 16.6a and 16.6b. Photograph and engraving of the well,
which lies deeper than the ground floor of the Lady Chapel,
and is located near the south wall of the ruins of the Lady Chapel.
My photo, 2017 (a). The engraving *The Holy Well of Glaston* is from *The
History of the Abbey of Glaston* by the Reverend Richard Warner (1826).
My photo of the engraving in the abbey museum (b).

wall of the original stone church, on which can be read two names: Jesus–Maria.[17] According to tradition, this Mary is Mary the mother, but the Dutch author Danielle van Dijk is convinced that the first church is dedicated to Mary Magdalene.[18] She wrote that Mary Magdalene had already visited Glastonbury with Joseph of Arimathea in 35 CE. Here, they both would have founded a Nazorean community. I do not exclude the possibility that both Mary the mother and Mary Magdalene visited Avalon. Mary the mother accompanied young Yeshua and her brother Joseph; Mary Magdalene could have visited Avalon after the crucifixion. Whether they are actually buried there is another matter, but they left traces. Joseph of Arimathea is said to have been buried right next to where Mary is said to be buried. That seems plausible given that he had settled here. The place where Joseph of Arimathea settled to bring the teachings of the Way to Britannia still is a place of great beauty.

Mary Magdalene in Avalon

According to legend, a hermitage or oratory dedicated to Mary Magdalene was said to have stood on a small hill southwest of the town close to Wearyall Hill. This mound is known today as Bride's Mound, because in 488 Saint Brigid of Kildare in Ireland took up residence in the Mary Magdalene hermitage and founded a small community here. The place attracts many pilgrims from Ireland and is still called Little Ireland or Beckery (*Beck* for "little" and *ery* for "Ireland") to this day.[19]

Glastonbury features a Mary Magdalene Chapel, located in Magdalene Street, directly adjacent to the abbey grounds. Once there was a hospital dedicated to Mary Magdalene in the thirteenth century, probably for poor women. When the elongated Mary Magdalene hospital with chapel fell into disrepair in the sixteenth century, it was divided into small houses for the poor. Mary Magdalene's name was

17. Jowett, *The Drama of the Lost Disciples*, 75.
18. van Dijk, *Maria Magdalena, de lady van Glastonbury and Iona*, 139.
19. Jones, *In the Nature of Avalon*, 15.

changed to Margaret, so then people spoke of Margaret's Chapel and the almshouses of Saint Margaret. Later this was reversed, and now people speak again of the Mary Magdalene Chapel and its almshouses.[20] The Magdalene Street ends at the Roman Catholic church of the hill, Mary of Glastonbury (fig. 16.7).

At the entrance to the extensive grounds of the Glastonbury Abbey is the chapel of Ireland's saint, Saint Patrick. Inside on the painted walls I encountered the arrival of Joseph of Arimathea with his floppy hat and traveling staff (fig. 13.0). There you will also find a painting of Mary Magdalene, depicted with the seven deadly sins fluttering around

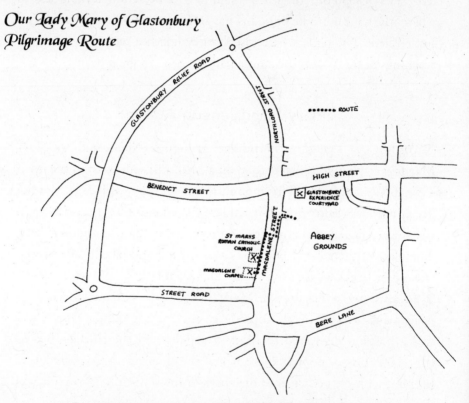

Figure 16.7. Map with Mary Magdalene Chapel on
Mary Magdalene Street, directly adjacent to the abbey grounds.
From Kathy Jones, *The Nature of Avalon*, 135.

---

20. Jones, *In the Nature of Avalon*, 137.

Figure 16.8. Mary Magdalene as queen, St. Patrick's chapel at the entrance to the abbey grounds. My photo from the same chapel.

her in bright ribbons like a wreath (fig. 16.8). It is the usual way of depicting her as a sinner and the saint of penance. Mary Magdalene, having conquered her sins, ascends into heaven; she is depicted tall and as a queen in the Chapel of Saint Patrick. Here, Mary Magdalene's ascension is portrayed exactly as it is in the Gospel of Miryam. Mary Magdalene left traces in Britannia's legends, in the landscape, and in the heritage of the priestesses of Avalon, who to this day keep the memory of her and the other Marys alive.

17

THE TRAIL OF
MARY MAGDALENE
FROM ANTIQUITY TO
THE MIDDLE AGES

In previous chapters, I followed the trail of Mary Magdalene in France and England in antiquity. The key question in this chapter is: Can a historical connection be made between Mary Magdalene in antiquity and the Cathars in the Middle Ages? After all, the Cathars preserved *The Gospel of the Beloved Companion* and released it in 2010. In that case, which historical links can be strung together in a living Magdalene tradition to form a chain that connects antiquity with the Middle Ages? Pillars that bridge ten centuries must be found; I found seven.

When the book *Holy Blood, Holy Grail* was published in 1982, I was impressed when reading it first in English and buying it later in the Dutch translation. Dan Brown drew a lot from this book when he wrote *The Da Vinci Code*. I met one of the authors of *Holy Blood, Holy Grail*, Sir Henry Lincoln, first in Rennes-le-Château and later at a musical symposium at a castle near Carcassonne. His silent and knowing presence made a deep impression on me. While rereading *Holy Blood, Holy Grail* for this book, I found seven links to connect the long chain of the French Magdalene tradition through the ages.

The authors of *Holy Blood, Holy Grail* had a working hypothesis: that descendants of Jesus and Mary Magdalene lived in the south of

France; in the early Middle Ages they married members of the royal family of the Merovingians; that made this royal house partly Hebrew. The three authors found numerous indications for this hypothesis in the early medieval Merovingian record, but no conclusive evidence; however, they did find convincing evidence for the continued existence of the House of David in the later Carolingian period.

In the story that follows, one area that plays a key role is Septimania in the south of France, which was part of the Roman province of Gallia Narbonensis beside the Mediterranean Sea. It contains the major cities of Carcassonne, Narbonne, and Béziers. To the east, the River Rhône separated Septimania from Provence. In the south the border was formed by the Pyrenees, and in the northwest the border was west of the River Garonne (fig. 17.1). The name Septimania may be derived from the Roman name for the city of Béziers or Colonia Julia Septimanorum Baeterrae. The veterans of the Roman seventh (in Latin *septem*) legion settled in Béziers, because they were allotted land there in return for their military service.[1]

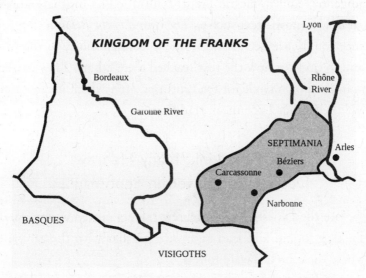

Figure 17.1. Septimania in the early Middle Ages, in 537 CE.
Author: Polylerus.

---

1. The *seven* (*septem* in Latin) in the name *Septi*mania may also be an allusion to the seven towns of the area: Béziers, Elne, Agde, Narbonne, Lodève, Maguelonne and Nîmes; see Wikipedia entry for "Septimania."

After the fall of the Western Roman Empire, Septimania came under the control of the Visigoths in 462. Many Hebrews lived in Roman Septimania, and they now connected with Visigoths, Merovingians, and Carolingians. Finally, under the Franks, Septimania developed into an independent Jewish kingdom for a century and a half; at the head of the area was a king who claimed to descend from David. It may be that the memory of Yeshua and Mary Magdalene, through their Hebrew descendants in Septimania, was passed on to the Visigoths, Merovingians, and Carolingians right down to modern times. It became world news when a priest, Bérenger Saunière, made discoveries about Mary Magdalene in 1917 in Rennes-le-Château. This theme, which stunned the world, was woven into the story of *Holy Blood, Holy Grail* and *The Da Vinci Code*.

My concern in what follows is to explain why the memory of Mary Magdalene remained alive in France more than elsewhere. My goal is to explain how the Cathars came to know about Mary Magdalene from the indigenous and domestic French culture. The ultimate aim is to find out how a gnostic text such as *The Gospel of the Beloved Companion* in France could have been handed down from antiquity to the Middle Ages and to find out how the text reached the Cathars. They passed the text on in their own circle for ten centuries, after which it came to us in 2010 via the Cathar Laconneau tradition.

The First Link: Many Hebrew Descendants Lived in Septimania

Long before the Christian era, Hebrew traders moved westward. They were looking for tin and lead and were familiar with the areas in the west where these metals were found.

Sources show that especially after 70 CE, the exodus from Palestine to the south of France grew into a true tidal wave.[2] As a result of several earlier diasporas, a large Hebrew population had settled in southern

2. Ralls, *Mary Magdalene*, 73.

France and northern Spain.³ The migrants spoke Greek and western Aramaic or Syriac. With Greek you could make yourself understood in the south of France, and there were numerous Hebrew Christian and Syriac Christian centers in southern France. For instance, many Hebrew people lived in comfortable villas in the area around Rennes-le-Château, where there were warm springs and fertile soil.⁴ This city in the south of France was part of an area traditionally called Septimania, which had a large Jewish population; a Jewish kingdom arose here in the eighth century.

The Second Link: Arius Denied that Jesus Was Divine

In the fourth century there was a theological battle between Arius (256–336), the founder of Arianism, and Athanasius, mouthpiece of orthodox Christianity. Arius was a presbyter or elder in Alexandria and was in contact with the great theologian Origen, and both were convinced of the subordination of the Son to the Father. Arius did not accept the Holy Trinity. He also emphasized the humanity of Jesus and denied his divinity; to him, Jesus was mortal and human and nothing more or less than an inspired wisdom teacher. This fits into the Hebrew-Christian framework that also emerged in the Apocrypha. The Visigoths were drawn to the theological ideas of Arius and Arianism did not clash with Judaism or with Islam, which emerged from Spain in the seventh century. Issa or Jesus in the Qur'an was mentioned thirty-five times as the Messenger of God or Messiah. He was nothing more than a mortal prophet, a forerunner of Muhammad and the spokesman of an

---

3. The three diasporas: 1) The first exodus occurred in 721 BCE when the Assyrians attacked the Northern Kingdom of Israel and deported ten tribes. 2) The second exodus followed around 600 BCE when the Babylonians besieged Jerusalem. A large part of the inhabitants of the southern kingdom of Judah fled their country; many took refuge in Alexandria, but people also went to the south of France and England. 3) The third exodus occurred when the Jews revolted against the Romans around 70 CE.
4. Rameijer, *Maria Magdalena in Frankrijk*, 74.

almighty god. The Qur'an, like Basilides and Mani, insisted that Jesus did not die on the cross: "They killed him not, nor did they crucify him, but they thought they did." Commentaries on the Qur'an say that a vicar died on the cross; Jesus hid in a niche in a wall and watched the crucifixion of a substitute.[5] Similar interpretations can also be found in Docetic texts in the Nag Hammadi library, which portray Jesus's death on the cross as a sham (chapter 12).

The Third Link: The Visigoths Sympathized with Arius and the Jews in Septimania

It is true that Arianism was condemned to heresy at the Council of Nicaea in 325, but in the early Middle Ages Arius's ideas still exerted an enormous influence on western Germanic rulers. By 360 Arianism had almost completely supplanted Roman Catholicism in Western Europe, despite the fact that Arianism had been declared a heresy. Why?

Numerous Germanic tribes adhered to Arianism starting in the fourth century.[6] Among them were the Visigoths; they sacked Rome in 480 and possibly took temple treasures, which the Romans took earlier from the destroyed temples in Jerusalem. The Visigoths likely brought these treasures to their capital in southern France, called Rhedae (later Rennes-le-Château). In Visigothic times, Rhedae was a large city with eight ramparts and a population of thirty thousand.

In Gaul or later France, the Arian Visigoths ruled in the fifth century. Under their influence, Arianism became the predominant form of Christianity in Spain, the Pyrenees, and southern France. The Hebrew people in France and elsewhere had little to fear from them.[7] The Visigothic nobility married into the Hebrew families who had been living in the south of France. In *Holy Blood, Holy Grail* you read that the

---

5. See chapter 12; Baigent, Leigh, and Lincoln, *Holy Blood, Holy Grail*, 386.
6. From the fourth century on, Goths, Suevi, Lombards, Alans, Vandals, Burgundians, Ostrogoths, and Visigoths were converted to Arianism.
7. Baigent, Leigh, and Lincoln, *Holy Blood, Holy Grail*, chap. 14, 383–89.

extended family of Jesus and Mary Magdalene may have mingled with the Visigoths, who also protected them against the Church of Rome.

The Fourth Link: The Merovingians Sympathized with the Visigoths and the Jews in Septimania

In the fifth century another Germanic tribe became powerful: the Franks, whose royal house consisted of the dynasty of the Merovingians (450–751). The early Merovingians were surrounded by neighbors such as the Visigoths and Burgundians who adhered to Arianism. The Merovingians, who sympathized with Arianism, married into the Visigothic Jewish nobility. In the Visigoth-Merovingian noble families and in the Merovingian royal house you will find many Hebrew names such as Samson, Miron the Levite, Solomon, and Elishachar (a variant of Eleazar and Lazarus). The father-in-law of the Merovingian King Dagobert I (603–638) was called Bera, a Semitic name. Bera's sister had married a member of the Levi family.[8] Numerous members of the Merovingian royal family continued to sympathize with Arianism. The Merovingians wore their hair long because they thought that their hair contained miraculous power. So did the Old Testament Nazarites, to whom Samson belonged.[9]

In the fifth century the Roman church was in a precarious situation. There was the Celtic Church in Britannia in the far west and there was Arianism, which was ubiquitous in continental Western Europe. The Roman Church had no primacy over the other churches. To survive, the Church needed a pillar of support. That strong arm was found in the person of the Merovingian King Clovis (466–511). He converted to Roman Catholicism and was baptized in 496. The pact between the Roman Catholic Church and Clovis formed the basis for the later power base of the Church over the other Christian groups in Western Europe. Clovis reigned from 481 to 511 and united the Frankish tribes under his

8. Baigent, Leigh, and Lincoln, *Holy Blood, Holy Grail*, chap. 14, 383–89.
9. Baigent, Leigh, and Lincoln, *Holy Blood, Holy Grail*, chap. 14, 383–87.

rule. The new Roman Catholic faith was imposed with the sword.

King Clovis's main enemies were the Visigoths; their empire extended on both sides of the Pyrenees. He defeated them in 507, forcing their retreat from Toulouse to Carcassonne and driving them from large parts of southern France. They settled at their last bastion, Rhedae (present-day Rennes-le-Château), in the area called the Razès.[10]

After the death of King Clovis his kingdom was divided among his sons, and this was the beginning of great division and strife. It weakened the position of the Merovingians and their mayors, the Carolingians, made grateful use of it.[11] When the Church saw the Merovingian dynasty weakening, it sought support from the Carolingians.

Not all Merovingian kings were divided and weak. The exception was the Merovingian King Dagobert II of Austrasia (651–679). After the death of Dagobert's father, the Carolingian mayor of the palace prevented Dagobert II from coming to the throne by kidnapping him as a child; Dagobert returned to France after many years and claimed the kingship. As a Merovingian, he joined the royal family of the Visigoths by marrying Giselle de Razès, daughter of the Count of Razès and niece of the Visigothic king. The marriage was celebrated in the stronghold of the Visigoths at Rhedae, namely in the church of Mary Magdalene, where the Mary Magdalene Church founded by the priest, Saunière, will later be located.[12]

Dagobert II and Giselle had a son, Sigebert IV, who also resided in Rhedae. In 674, Dagobert II was proclaimed king. He would have amassed great treasure in Rhedae to reconquer Aquitaine, which had separated from the Merovingians forty years before and had existed since that time as an independent principality. Three years later, Dagobert II was killed by a servant while hunting (commissioned by the mayor of the palace, Pépin the Fat, with the knowledge of the Roman Catholic Church).

---

10. Baigent, Leigh, and Lincoln, *Holy Blood, Holy Grail*, chap. 9, 242–45, 247.
11. Baigent, Leigh, and Lincoln, *Holy Blood, Holy Grail*, chap. 9, 245–46.
12. Baigent, Leigh, and Lincoln, *Holy Blood, Holy Grail*, chap. 9, 247, 250.

The Carolingians began their advance. The last Merovingian king, Childeric III, was deposed in 751 by the Carolingian Pépin the Short. According to a dubious story, Sigebert IV (Dagobert II's child) was killed in a hunting accident at the age of two, two years before his father's death. But this story seems to be wrong because Sigebert IV survived his murdered father, going on to have Merovingian children. The theory is that the family of the Merovingians—under the new royal house of the Carolingians—had secretly propagated.[13] According to the authors of *Holy Blood, Holy Grail* it would be this royal blood that flowed through the veins of the Jewish kings of the later Jewish kingdom Septimania.

The Fifth Link: The Arrival of the Moors and Jewish Cooperation with the Carolingians

In 711, the situation of the Jews in Septimania and northeastern Spain had deteriorated sharply. Dagobert II was murdered in 679, and his descendants hid in the area around Rennes-le-Château in the Razès. The Merovingians still nominally sat on the throne, but in reality power lay in the hands of their mayors, the Carolingians, who with the support of Rome were ready to establish their own royal dynasty. When Visigothic Spain was conquered by the Moors in 711, the Jews welcomed the Moorish invaders.[14] The Jews led a flourishing existence in Spain under Mohammedan rule. The Moors treated them with courtesy and entrusted them with the administration of conquered cities. In the early eighth century, the Moors entered Septimania; it was in Muslim hands from 720 to 759 while the last Merovingian descendants hid in the area around Rennes-le-Château in the Razès.[15]

The Moorish advance was stopped by Charles Martel, a mayor of the palace and the grandfather of Charlemagne. Charles Martel's son, Pépin,

---

13. Baigent, Leigh, and Lincoln, *Holy Blood, Holy Grail*, chap. 9, 257.
14. Baigent, Leigh, and Lincoln, *Holy Blood, Holy Grail*, chap. 14, 389–90.
15. Baigent, Leigh, and Lincoln, *Holy Blood, Holy Grail*, chap. 14, 389–91.

made agreements with the local aristocracy of Septimania around 752. He got the area under control, except for the city of Narbonne. In 759 Pépin concluded a pact with the Jewish population of Narbonne; they opened the gates for the besiegers and helped him against the Moors. In return, he allowed the Jews in Septimania to form their own principality with their own king. In 768, Septimania became a Jewish principality; it was subject in name only to Pépin but was in fact independent.

In summary, the links between antiquity and the Middle Ages run from the Hebrews in Septimania in southeastern France via the Visigoths, Merovingians, and Carolingians to a Hebrew principality in Septimania. At its head was a King of the Jews who descended from the house of David. He was both Merovingian and of Jewish blood.[16] Contemporaries of the seventh and eighth centuries took this lineage completely seriously and showed great respect.

The Sixth Link: Septimania Transformed into a Jewish Principality

The ruler of Septimania was hereafter officially referred to as King of the Jews. It was Theodoric or Thierry, the father of William of Gellone. Both Pépin and the Caliph of Baghdad acknowledged that he was from "the seed of the royal house of David."[17] Septimania flourished in the following years, partly because it received estates from the Carolingian kings. Theodoric or Thierry married Alda, Pépin's sister, to strengthen the alliance between his royal family in Septimania and the Carolingians.[18]

Thierry's son, William of Gellone, was also seen in Septimania as the King of the Jews. He was also Count of Barcelona, Toulouse, Auvergne, and Razès. Like his father, he was a Merovingian and a Jew of royal blood. Recent research has conclusively shown that William

---

16. Baigent, Leigh, and Lincoln, *Holy Blood, Holy Grail,* chap. 14, 393.
17. Baigent, Leigh, and Lincoln, *Holy Blood, Holy Grail,* chap. 14, 394.
18. Baigent, Leigh, and Lincoln, *Holy Blood, Holy Grail,* chap.14, 393–94.

was Jewish. In chivalric novels in which he appears as Willem, Prince of Orange, he spoke fluent Hebrew and Arabic. His coat of arms featured the lion of Judah, the symbol of the house of David and Jesus, and he observed the Sabbath and Jewish holidays, even during his campaigns.[19]

William was one of Charlemagne's knights. When Charlemagne's son, Louis, was installed as Emperor, William placed the crown on his head. Louis is believed to have said: "Lord Willem . . . your family has exalted mine." It is clear to Louis that William belonged to an old family, an older and more elevated family than his own.[20]

William was more than a soldier. In 792, he founded an academy and attracted scholars and set up a famous library. Gellone became an important center for Jewish studies. In 806, he retired from active life and concentrated on the academy, and Gellone became a place where Mary Magdalene was venerated. William died there in 812. Later the academy was turned into a monastery, Saint-Guilhelm-le-Désert.

The Seventh Link: The First Cathars

Mary Magdalene is still called The Lady of Light or La Dame de la Lumière in France to this day, and she is venerated in a special way in Provence to this day, much more often and more intensely than Mary, Yeshua's mother. Numerous medieval churches, chapels, paintings, sculptures, and writings testify to the deep impression she must have left on these regions. Why is this? Is there an explanation for it?

Descendants of Yeshua and Mary Magdalene may have lived on during the reign of the Merovingians and Visigoths. Perhaps the descendants of the Merovingians did not die out after the takeover of power by the Carolingians. Perhaps they lived on underground and passed on the Magdalene legacy. All this took place in the deep south of France at the foot of the Pyrenees, where large groups of Hebrew refugees once

19. Baigent, Leigh, and Lincoln, *Holy Blood, Holy Grail*, chap.14, 394, 396.
20. Baigent, Leigh, and Lincoln, *Holy Blood, Holy Grail*, chap.14, 394.

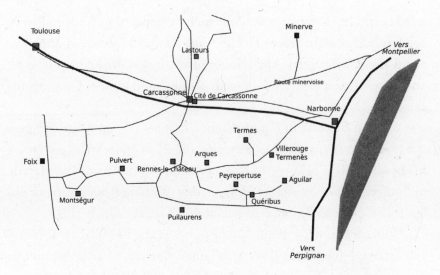

Figure 17.2. Map of the routes of Cathars in medieval France.
Source: Made with Inkscape.

settled in ancient Septimania; perhaps this explains why the love for
Mary Magdalene is widely supported in France.

The medieval Cathars stand in an older indigenous tradition known
and preserved in the south of France; it conveys the true history of
Mary Magdalene and her family, said to have lived in Rhedae (Rennes-
le-Château). As we learned earlier, Rhedae later became the capital of
the Visigothic Empire and the Merovingian monarchs who took refuge
there had a special relationship with Mary Magdalene. For example,
King Clovis visited the tomb of Mary Magdalene, and many French
monarchs and dignitaries followed him. They saw Mary Magdalene as
the patroness of France.

As mentioned earlier, Bérenger Saunière found secret documents
in 1917 in the Saint Mary Magdalene Church of Rennes-le-Château,
which was built on the foundations of the Visigothic castle. The documents traced the ancestral lineage of the Merovingian King Dagobert II.
Saunière was also said to have found a treasure trove of money here that
enabled him to finance major construction projects in the church, rec-
tory, gardens, and the Magdala tower. Books like *Holy Blood, Holy Grail*

and *The Da Vinci Code* became world bestsellers, which led the world to take a new interest in Mary Magdalene as the wife of Yeshua and, according to *The Gospel of the Beloved Companion*, his immediate successor.

The first Cathars appeared in the tenth century, and by the eleventh century, more and more Cathar centers where apocrypha circulated, were discovered. Increasing numbers of authors view the Cathars as an indigenous French movement. For example, in a previous book I wrote that French authors argue in favor of considering the early forerunners of what would swell into a powerful Cathar stream after 1150 as an independent phenomenon, without involving influence from the Balkans. According to René Nelli, the early moderate Cathars were an independent French phenomenon and originated in the south of France. The authors of *Holy Blood, Holy Grail* repeated that the Cathars were of indigenous origin and are moderately dualistic, so they did not come from Bulgaria or Italy, where later a more extreme dualism originated.[21]

Did Catharism Become More Dualistic in the Twelfth Century?

In 1167, a Cathar synod consisting of representatives from northern France, northern Italy, and the Balkans, met in Saint-Félix-Lauragais east of Toulouse. Bishop Nicetas of the Bogomils of Constantinople presided. The moderate dualism would then have developed into an extreme dualism in France.

The German researcher Konrad Dietzfelbinger argues that two texts regarded by inquisitors as the French Cathar "bible" do not come from France but from neighboring countries.

- The *Interrogatio Johannis* or *The Interrogation of John* is a text in which John questions Jesus. The text was confiscated by inquisitors in Carcassonne. This document was not written by the

---

21. Baigent, Leigh, and Lincoln, *Holy Blood, Holy Grail*, chap. 2, 57–58; chap. 14, 383–85.

French Cathars themselves; it probably came from the Bogomils of Bulgaria and reached France via the Italian Cathars.[22]

- The *Liber de duobus prinicipiis* was a text that came from Italy.[23] The question was to what extent both writings were benchmarks for French Catharism. They dealt with the relationship between God and Satan, with Satan as the creator of a sinful world from which man had to escape through knowledge and insight.

In philosophy, we distinguish between monism and dualism. A monist thinks from unity (*monos*). A dualist sees two independent, often contradictory and therefore opposing poles. The original gnosis assumes Oneness between the worlds of spirit and matter; matter being the condensed and dense expression of spirit. Despite the monistic tendencies, gnosis and gnosticism and later Catharism are described by the Church and science as being dualistic. Until the high Middle Ages, Cathar Christians were pursued and stigmatized by ecclesiastical theologians; many were condemned as "dualists" and "devil worshippers" and were wrongly burned at the stake.[24]

The French Cathars

The collective term *Cathars* ("pure ones") was not applied to the movement by their opponents until much later. The French word *parfait* (from the Latin word *perfectus* or "a perfect one") was also not an authentic Cathar word. It came from the Inquisition and was used to indicate the Cathar clergy comprising both men and women.

The French Cathars were known in their time for their honest, pure, and sober way of life. They called themselves the *boni Christiani* or the "good Christians"; in France people commonly called them the *bonshommes* and *bonnesfemmes*, "the good men and women," because their priests and teachers included both. They saw themselves as

22. Dietzfelbinger, *Mysteriescholen* (*Mystery Schools*), 236.
23. Dietzfelbinger, *Mysteriescholen*, 236.
24. Dietzfelbinger, *Mysteriescholen*, 246–47.

friends of God and considered their Church to be the true church of God. Around the year 1000, a mystery school flourished in southern France. The Cathars wished to vitalize and revive old forms of original Christianity and gnosis. For a Cathar, the Gospel was a mystery writing indicating a path of initiation. Man and his world were in an imperfect condition; there was a perfect Divine World to which a human being belonged with a higher part of his being.[25]

These are many different Cathar groups under many names. They are also called Albigenses, after the city of Albi, which was one of their centers where they were condemned by the Church in 1165. Albigenses and Cathars were collective names but the group as a whole involved many separate subgroups led by independent leaders. Much of the information about them comes from the Inquisition. The authors of *Holy Blood, Holy Grail* stated, "It is therefore impossible to fully describe the Cathar way of thinking."[26] They believed in reincarnation and recognized the feminine principle in religion as evidenced by both sexes serving as preachers and teachers. They emphasized direct and personal knowledge and mystical experience or gnosis. The emphasis was on direct personal contact with God.

Following the Gospel, Cathar priests lived very simply and soberly in contrast to the splendor in which Catholic prelates lived. Around 1200, large parts of the population of southern France converted to Catharism. The Church decided to act. (How that happens, you can read in the next chapter.)

In this chapter, we have followed seven links connecting antiquity with the Middle Ages, and it is quite possible, even plausible, that the memory of Mary Magdalene's ministry and teachings in the south of France has been kept alive through these links. The descendants of her extended family may have passed this legacy down to the Middle Ages in southern France through oral and written transmission—but while possible, it has not yet been proven.

---

25. Dietzfelbinger, *Mysteriescholen*, 234.
26. Baigent, Leigh, and Lincoln, *Holy Blood, Holy Grail*, chap. 2, 52.

18
MARY MAGDALENE AND
THE CATHARS

The Magdalene tradition was common to all people living in France, including French ecclesiastical officials, until about 1200. It was not an apocryphal tradition; it was a canonical tradition recognized by the French Church. This chapter, including regression memories and channeled information, shows that this Magdalene tradition had been completely absorbed into the twelfth-century monastic culture of central and southern France. Again, the question is: To what extent do the historical sources support these memories?

The Regression Memories of Pierre:
Monk of Cîteaux

In regression, Cathie Welchman remembered being a monk named Pierre from the twelfth century, who had traveled with his father, Joseph, to the south of France. Joseph was the Abbot of Cîteaux Abbey situated south of the city of Dijon, and he and his son had traveled together to the town of Mirepoix in southern France, where the monastery also owned lands southwest of Carçassonne.

The fortified town of Mirepoix was a Cathar stronghold in the Pamiers district of the Ariège.[1] Despite the fact that they felt sympathy

---

1. For Mirepoix as Cathar stronghold, see "Mirepoix," at the Terres Cathares website.

for the French Cathars, they were required to report to the Pope about them, otherwise the monastery's supply of money would be cut off. They felt very uncomfortable with this. On the one hand, they shared a love of Mary Magdalene with the Cathars and realized that they, like the Cathars, spoke French and were French. On the other hand, they feared the power of the Roman Catholic Church that wanted to dominate France, and they knew they depended upon it. Joseph and Pierre's reconnaissance on the Cathars preceded the first Albigensian Crusade (1209–1229), which began with the massacre in Béziers on July 22, 1209—Mary Magdalene's name day.

Father and son were on their way to visit the Cathars, and the next passage can be dated somewhere between 1140 and 1209. It seems likely to me that it took place around 1150, because by then the Cathar movement in the south of France had grown into a large and popular movement, and the southern French population's love for both the Cathars and the indigenous Jewish population was a thorn in the side of the Church of Rome. Pierre said,

We are going to meet the Cathars in a fortified city where most inhabitants are living openly as Cathars. They feel a sense of comradeship with us because we are all French and we're both afraid of what the Roman Pope is doing. We can only think he means to take over France, he wants the power. In our monastery we follow the Way, the Way of Mary Magdalene and the Chalice. . . . The Chalice of Love . . . to do to others what you would expect to be done to yourself . . . to dip into the Chalice is to link with all other healers, to feel as they feel, to expect to receive what you give out.

That is why we are allowed to be married. It is part of the love of a human being. The Roman Catholic Church does not allow their priests to be married. It's a form of control. They must give all of their life to the Church and not have anything else interfere. Here in France it has been that way for centuries for the monasteries, trying to dispense love to the people and teach the people how to be—we're peacemakers. People come to us to receive compassion and forgiveness. We help them to understand how they should forgive and live with their neighbors in peace. This teaching comes from the Lord. The Lord sent his emissaries out, we received the word and we have been keeping it. It was not the Pope. The Pope is telling us to subjugate anyone who doesn't do what he says is right. That's not the path of peace. The Cathars are different from us, but they are still working to live in peace with each other. They do have some strange religious practices that we don't have. Many of them take a vow to be celibate towards the end of their lives, but some take it earlier and they remain celibate. We don't do that.

It was all tied in with the Magdalene. I am not sure if the Cathars had some link with the Magdalene. It's we who remind ourselves how the Magdalene came. She traveled with her kith and kind . . . sailing by sea . . . to our shores. She brought goodness to our land, and it's all destroyed by this evil man, the Pope. You can feel it, it's tangible. He's blowing black clouds over a region that's been peaceful for many centuries. We've grown in love and suddenly it's all changing. We have a history that goes right back, and the Pope is even destroying our history. He's burning books for no good reason.[2]

When asked whether Pierre knew where the Magdalene came from, he replied:

She came from the Holy Land to spread the word of the Lord to us. We've written about it in books and they've been destroyed. And we're trying to hide the knowledge and it feels as if it's all coming to an end and we won't be able to live this way of life.

We had never met the Cathars before because we never needed to, but now we are travelling because the papal edicts have decreed

---

2. Wilson and Prentis, *Power of the Magdalene*, 121–22.

this. We travel and report back what the Cathars are doing. We know underneath that we'd rather side with the Cathars, but we must report back. We must write what we find. If we don't, we'll be in trouble. It feels very dangerous, this path. We're on nobody's side. We can't be on the Cathars' side even though we'd prefer it, because we're doing work for the Pope. We don't want to be on the side of the Pope because he feels so dangerous. He feels as if he's come straight from the devil. But of course, he's paying for the monastery to exist. The coffers in Rome are filled up and if you don't do as asked by your master . . . you'll cease to exist. And we must protect the Way of the Magdalene somehow. The message of the Magdalene was to share bread with everyone you love, share. Remember, in love, to understand your neighbor. The best way to understand your neighbor is to break the bread and eat with them, share.[3]

Prentis asked: "And also the healing was important?"

Pierre replied: "We're getting to the point now where we can't even give people healing. The Pope has set up apothecaries and they give a potion which doesn't do anything for the people."[4]

Commenting on this passage, Stuart Wilson quite rightly, I think, observed the following: "These passages show that there were 'sharp differences' between the French Catholics and the Italian Catholics. In France, Mary Magdalene was seen as an important teacher and leader (in Italy this was Mother Mary). The Pope was carrying out a policy of centralization over the French Church, with Mary Magdalene increasingly fading into the background."[5]

In the following sections, I search for historical background information, asking the question: To what extent are these old memories supported by historical evidence?

3. Wilson and Prentis, *Power of the Magdalene*, 123.
4. Wilson and Prentis, *Power of the Magdalene*, 123–24.
5. Wilson and Prentis, *Power of the Magdalene*, 124.

A LADYLIKE INTERLUDE:
WAS NOTRE DAME MARY THE MOTHER OR
MARY MAGDALENE THE BRIDE?

Joseph, Pierre's father, was said to be Abbot of the Cîteaux Monastery. Founded in 1098, this monastery was the headquarters and center of the Cistercian order. The settlements of the order sprang up like mushrooms in the twelfth century, and by the beginning of the thirteenth century the order had five hundred abbeys. I have not come across Joseph's name in the list of abbots of Cîteaux; he may have been abbot of one of the smaller monasteries that fell under the order.[6] The historical sources do provide a lot of information about Arnaud or Arnoldus, the seventeenth Abbot of Cîteaux. He played an extremely negative role in the prelude to the Albigensian War, which started in 1209. Before the crusade, he traveled in splendor through southern France to win back the Cathars for the Church. Later, this Arnaud of Cîteaux became head of the crusade army. He hated the Cathars and Jews. Joseph, Pierre's father, seems to have been of a completely different caliber.

Robert of Molesme, the first Abbot of Cîteaux, showed a great love for Notre Dame, from whom he was said to have received a white hood. In 1115, Bernard of Clairvaux (1090–1153) joined the order and breathed new life into the ailing order by transforming it into a multinational religious order comprising hundreds of abbeys from the far West to Russia. Each abbey was dedicated to Notre Dame or Our Lady. The big question is: Who was this Lady?

Bernard was born near Dijon into a noble family in the village of Fontaine-lès-Dijon, which once had a Black Madonna. As a little boy, Bernard received three drops of milk from her

---

6. Wikipedia has a list of Abbots at Cîteaux Abbey.

Figure 18.1. Saint Bernard of Clairvaux. Engraving. The Black Madonna
of Fontaine-lès-Dijon (the native village of Bernard of Clairvaux)
sprinkles water on Bernard's lips; she is depicted here as
La Virgen de la Leche y Buen Parto in the so-called Milk Wonder.
Author: Wellcome Library, London.

breast, the so-called "milk miracle" (fig. 18.1). This would explain the great deeds he later performed. The city of Dijon itself has to this day a famous dark Lady called *Notre-Dame de Bon-Espoir* or The Black Lady of Good Hope (fig. 18.2).

In 1128, Bernard of Clairvaux wrote the regulations for the order of the Knights Templar; Mary Magdalene received special attention in this. Separately, Bernard wrote hymns and sermons about Mary. He compiled about 280 sermons about the Song of Songs, which dealt with the marriage between Solomon and his bride, a lady, black and beautiful.[7] In other words, he showed

---

7. Sg. 1:5–6: The Lady says: "I am black and beautiful (*nigra sum et formosa*)."

Figure 18.2. *Notre-Dame de Bon-Espoir* or "The Black Lady of Good Hope" from Dijon. She is one of the oldest Romanesque images from around 1150 and is one of the most beloved dark mothers from Burgundy. She is pregnant and is said to have been depicted with a child. She lost her child, hands, and feet during the French Revolution. She is made of oak and measures eighty-four centimeters long.

great interest in the spiritual marriage as described in the Song of Songs, a wedding song. Bernard's heart overflowed with love for Mary the Bride. He has been called the first troubadour of the Madonna or Notre Dame. Because of the love for the lady and the spiritual marriage that Bernard propagated, there was a run of women on the Cistercian monasteries. Apparently, women and men were living in so-called double monasteries, a long-established tradition in Celtic Christianity in Western and Central Europe and proven to still be present in the west of France around 1000 CE.[8] At a certain point the abbots could no longer cope with the influx and founded separate women's monasteries.[9]

---

8. The father of itinerant preacher Robert of Arbrissel (1045–1117), from Brittany, was a parish priest who is said to have converted prostitutes in his following.
9. Southern, *Western Society and the Church in the Middle Ages*, 315: for the negative attitude of the Cistercian monks toward the Cistercian nuns.

Traditional commentaries state that this sometimes-dark lady was the Virgin Mary. Others such as Lynn Pickett and Margaret Starbird were convinced that the Black Lady and Bride of the Song of Songs was Mary Magdalene.[10] Between 1050 and 1300 eighty cathedrals dedicated to Notre Dame appeared in France. In addition, five hundred larger churches were dedicated to Notre Dame in cities, and thousands of smaller parish churches dedicated to Notre Dame were built in the countryside. Jaap Rameijer believes that all cathedrals dedicated to Notre Dame and built at the initiative of the Cistercians and Knights Templar were in fact dedicated to Mary Magdalene, meaning Notre Dame concerns Mary Magdalene and not Mary the mother.[11] For years, I have not been able to answer the question of which Mary it was. But now that I survey the historical sources from antiquity to the Middle Ages, I am sure: once you know how entrenched Mary Magdalene was in the French collective memory, the question of who was hiding in Notre Dame is no longer difficult to answer. It is Mary Magdalene.

A CELIBATARIAN INTERLUDE: WHEN DID CELIBACY BECOME MANDATORY?

Celibacy was not compulsory in France yet. How else could Joseph, abbot of one of the monasteries within the Cîteaux order, have a son who was a monk (Pierre)? In Europe, the Church had been organized at a local and regional level until 1000; often clergy were under the protection of the aristocracy. Priests were married and the office of priest was passed down

10. Starbird, *The Woman with the Alabaster Jar*, 32; Picknett, *Mary Magdalene: Christianity's Hidden Goddess*, 134.
11. Rameijer, *Maria Magdalena in Frankrijk*, 68, 254.

from father to son in the family. In the Gregorian reform that spread in the eleventh century, opponents of the clerical marriage intended to declare this hereditary office invalid because money from the church community should go to the church and not to a priest's family. Furthermore, a priest needed to give his attention to the church and not to his wife and children. When priestly marriage was abolished, the priest's property no longer belonged to his family, but to the Church.

Begun in 1045, the Gregorian Reformation, named after the reformist Pope Gregory VII (1073–1085), spread like wildfire across Europe. The aim of every good Christian was to lead an apostolic and ascetic life in imitation of a supposedly poor, chaste, and unworldly Jesus Christ. In the medieval world view, Christ would have wandered around as a poor and itinerant preacher without a wife, family, possessions, or permanent residence. A good medieval Christian needed to imitate this ascetic Christ. A new impetus of purity was sweeping over Europe; numerous new ascetic monastic orders were founded. The papacy went along with this development; it wanted to purify, reform, and centralize, including in France. Reform and centralization of power from Rome ultimately made the Church of Rome grow into a secularized world power, and the reformed papacy embarked on a crusade against abuses including clerical marriage.

Finally, the prohibition of clerical marriage and the obligation to clerical celibacy were promulgated as canon law at the First and Second Lateran Councils in 1123 and 1139. Compulsory clerical celibacy was only introduced during the twelfth century. It was not followed immediately everywhere due to great resistance. It was therefore not surprising to encounter married priests in France in the twelfth century. What seems a bit strange, though, is that an abbot had a son with whom he traveled openly; but again there was this age-old Celtic tradition in western Europe of double monasteries and the cohabitation of male clergy and their female assistants—of nuns and monks, female hermits and male hermits—living together.[12]

The historical sources clarify that the love and respect for Mary Magdalene or Notre Dame was widely supported in France. The historical sources did not exclude that the veneration of Mary Magdalene in France was part of the canonical tradition in churches and monasteries. On the contrary, they seem to support this right up to the tipping point: the crusade against the Cathars that began in 1209.

How to Evaluate the Cathar Movement? The Alternative Sources

To Prentis's final question of how to value the Cathar movement as a whole, Alariel replied:

The Cathars provided in Languedoc an alternative focus of Christian culture, a focus which respected the Earth and honored the Sacred Feminine. However, it was not as pure a form of Christianity as the Gnostic movement, and some aspects of Cathar theology can only be described as eccentric. Yet, for the most part, the Cathars did live in Unconditional Love and Christian Harmony, at a time when this was almost unknown anymore else in Europe.[13]

Alariel explained further, and I summarize, Mary Magdalena brought Gnostic knowledge to France, forbidden knowledge, which

---

12. This was referred to as *syneisaktism* (spiritual marriage), a form of cohabitation of couples of a male and a female hermit, whereby the female ascetics are also called *virgines subintroductae*. In the sixth century, Breton priests appeared as itinerant preachers and took spiritual women with them as spiritual sisters or *conhospitae*; these assisted them in the performance of their priestly duties. Later, in the eleventh and twelfth centuries, itinerant preacher and Breton monk Robert van Arbrissel and many others followed in their footsteps.

13. Wilson and Prentis, *Power of the Magdalene*, 125.

gave people insight and empowered them. But the Church had no interest in enlightened and empowered people who questioned the authority
of the Church as a mediator between God and man. When the last
Gnostic died in the third century, most of this secret knowledge died
with him. It continued to be taught in Mary's extended family in the
south of France, but so few teachers remained that they were unable to
protect the purity of the transmission.[14]

This led to mistaken ideas, like the dual creation, creeping into the
tradition by the time of the Cathars. (In a dualistic worldview, the good
God was opposed to a bad devil on the same level, and the earthly creation was the work of the devil.) The Cathars retained the practice of the
Way, but by then their knowledge of Mary's philosophy had largely been
lost. (Mary's philosophy of Oneness implied that the earthly creation was
also part of God's creation.) According to Alariel: "The Essenes were a
Pure Transmission and so was Mary Magdalene's teaching in her mystery
school, but when the last Essene community was destroyed and the last
Gnostic died, that time of Pure Transmission came to an end."[15]

That is why this process of recovery is so important. It restores the
Pure Transmission of Mary Magdalene's teaching to the world. At the
end of the planetary cycle, the veils of forgetting are dissolving and
knowledge lost for many centuries is being restored. This teaching is
one of the great jewels of human achievement, and with humanity moving through Transition, it is entirely right and just that access should be
restored to you, as it may prove helpful at this time.[16]

The Crusade Against the Cathars and
Mary Magdalene: The Historical Sources

Around 1200, large parts of the Languedoc population were converted
to Catharism. In the thirteenth century the Church decided to act by

---

14. Wilson and Prentis, *The Magdalene Version*, 76.
15. Wilson and Prentis, *The Magdalene Version*, 77.
16. Wilson and Prentis, *The Magdalene Version*, 77.

sending a delegation to try to win back the renegades. The group consisted of Arnaud Amalric, the seventeenth Abbot of Cîteaux and head of the Cistercian order between 1202–2012, and a number of monks from Citeaux. The papal delegation passed through the country with pomp and show. They preached and called for penance, but they only reaped distrust and further estrangement.[17]

A fanatical Spanish monk, Dominic de Guzmán, later joined the delegation. He understood that only a modest, austere, and evangelical life, as shown by the Cathars, could impress the population, so he traveled barefoot through the country, but this tactic also had little effect. In 1216 he founded the Dominican order, which was aimed at converting the Cathars. From this came the Inquisition. The Cathars were not the only victims; the indigenous Jewish population, which had so far enjoyed the protection of the southern French nobility, also suffered.[18]

After the failure of the papal delegation, the pope appointed Arnaud as papal legate and inquisitor in 1204. Together with Pierre de Castelneau, he was sent to Septimania in Languedoc to suppress the Cathars. The northern noblemen had long looked greedily at the rich goods of the southern French nobility and the prosperous cities in the south. They became the stormtroopers of the Church of Rome, which took advantage of their greed and jealousy. It was a situation ripe for provocation.

On January 14, 1208, Pierre de Castelneau was murdered in Languedoc. The perpetrators likely had no ties to the Cathars, but the Cathars were accused of it anyway. At the instigation of Arnaud, Pope Innocent III called for a crusade. Arnaud headed an army of about thirty thousand knights and soldiers that settled in Languedoc in 1209.

The Albigensian Crusade, named after the French town of Albi, began on July 22, 1209—Mary Magdalene's name day. It started with the massacre in Béziers, where a large group of Cathars lived alongside Catholics and Jews. According to the account of the Cistercian writer

---

17. Dietzfelbinger, *Mysteriescholen*, 252.
18. Baigent, Leigh, and Lincoln, *Holy Blood, Holy Grail*, chap. 2, 55–56.

Caearius of Heisterbach, when a crusader asked Arnaud how to distinguish the heretics from the true believers, he is alleged to have said, "Kill all. God recognizes his own people."

In a letter to the Pope from 1209, Arnaud of Cîteaux wrote that about twenty thousand people died by the sword, regardless of caste, age, and sex. After this, the city was sacked and burned, so that "the wrath of God rages in it in a wondrous manner." According to another report from Pierre de Vaux-de-Cernay, seven thousand people had taken shelter in the Mary Magdalene Church; the doors were kicked open and all were killed. Both numbers are believed to be exaggerated.[19] The Jewish records were specific about how many members of Beziers' thriving Jewish community were exterminated: two hundred people were killed, and a large number of others were captured.[20]

After Béziers, the crusaders marched through the entire Languedoc and assaulted Narbonne, Carcassonne, and Toulouse. In 1210, when the city of Minerve capitulated, there were 140 Cathars burned at the stake, and Arnaud once again played an irreconcilable role. The killing was on such a grand scale that it may have been the first genocide in European history. The crusade army destroyed, burnt, and brutally wreaked havoc in the south of France. An entire culture was destroyed.

The Siege of the Cathar Fortress on Montségur: The Historical Sources

In 1243, in the last throes of the Cathar resistance, the fortified castle of Montségur was besieged. It was located on a rock at an altitude of 1,200 meters in the foothills of the Pyrenees, eighty kilometers south of Carcassonne. One of the centers of southern French Catharism, it was

---

19. The population of the town is estimated to have been between 10,000 and 14,500 people.
20. Baigent, Leigh, and Lincoln, *Holy Blood, Holy Grail*, chap. 2, 55–57; Gottheil and Enelow, "Arnold of Cîteaux" on the Jewish Encyclopedia website.

Figure 18.3. The fortified castle, the stronghold Montségur.

home to the Cathar Bishop Guilhabert de Castres and the seat of the Cathar Church since 1233 (fig. 18.3).

For ten months the fortified castle was besieged by about ten thousand men. They surrounded the entire mountain and tried to starve out the defenders. Some besiegers were sympathetic to the Cathars. The besieged could slip through the gaps in the lines and bring in fresh supplies.

According to rumors, a treasure was kept in the castle. After the surrender, nothing was found. Earlier in January 1244 and almost three months before the fortress surrendered in March, two parfaits escaped. According to reliable sources, they took most of the stock of gold, silver, and coins to a fortified cave in the mountains and from there to a castle. The treasure has since disappeared.[21] Rennes-le-Château is half a day on horseback from Montségur.

---

21. Baigent, Leigh, and Lincoln, *Holy Blood, Holy Grail*, chap. 2, 58.

334 ⊛ In the Footsteps of Mary Magdalene

On March 1, Montségur finally capitulated. There were less than four hundred defenders left; between 150 and 180 of them were *parfaits*, and the rest were knights, squires, and family. Conditions were imposed upon the capitulation. The defenders asked for a two-week reflection period. The attackers agreed to a two-week truce. Hostages were extradited, and it was agreed that they would be killed if someone tried to escape from the fortress. The fact that the besieged at the top of the fortress asked for a two-week delay may have to do with the arrival of spring and the celebration of the spring festival in freedom before finally surrendering.

The Cathars did not celebrate Easter because they doubted the importance of the crucifixion. But on March 14, the day before the ceasefire ended, a festivity took place from which an important "treasure" was later smuggled away. During that festivity twenty-one occupiers—six women and about fifteen soldiers—voluntarily received the *consolamentum* or the baptism in Spirit. With this they also became parfaits and condemned themselves to a certain death. During this ceremony the Gospel of John—or much more likely its source document, *The Gospel of the Beloved Companion*—was placed on the head of the one who received the Spirit. The initiated Cathars present would place their hands on the Gospel while it was on the head of the one receiving the consolamentum. Then the beginning of this Gospel was solemnly read aloud.[22] The Gospel was necessary in order to transmit the blessings of the Feminine Spirit, so very present in *The Gospel of the Beloved Companion* and in the Cathar Church as the Church of the (Feminine) Spirit. That could be the reason that it was absolutely necessary to keep the priceless gospel in the castle during this perilous situation.

The truce expired on March 15, 1244. At early dawn approximately two hundred parfaits were taken down the mountain. None of them recanted the faith. There was no time to erect individual pyres; they were herded into a palisade and burned together. The rest of the

22. Dietzfelbinger, *Mysteriescholen*, 245.

garrison watched from the castle. They were told that any attempt to escape would end in death for all of them and the hostages.[23]

What Treasure Was Smuggled Out of Montségur? The Cathar and Alternative Sources

Still, the garrison had allowed four parfaits to hide. Overnight on March 15, with the knowledge and consent of the garrison and accompanied by a guide, these four undertook a daring escape.[24] They descended along ropes down the steep western slope of the mountain, sometimes dropping more than a hundred meters in one go. With this action they endangered many lives. Why? It couldn't have been a monetary treasure; that was smuggled away months earlier. Did the treasure consist of manuscripts? Why weren't they taken away months earlier? Why hold them back in the fortress until the last, most dangerous moment? What did that treasure consist of?

Prentis asked Alariel what the small group who managed to escape were carrying with them; after all, it was said that they were bringing something of great value to safety. Alariel replied, "That was their most precious scroll dating from the time of the gnostics, demonstrating that there was a direct link between the gnostic tradition and the Cathars. This scroll was ancient by that time, and it was much revered by the Cathars."[25]

The Cathar Laconneau community is convinced that this was not a treasure of precious metal but a treasure of precious word. The treasure that was carried away at the eleventh hour concerned their beloved text, *The Gospel of the Beloved Companion*. The text was secured and from that day on only used internally.

The vine will bloom again; on the third page of the book *The Gospel of the Beloved Companion* is the first dedication to the Cathars:

---

23. Baigent, Leigh, and Lincoln, *Holy Blood, Holy Grail,* chap. 2, 59–60.
24. Baigent, Leigh, and Lincoln, *Holy Blood, Holy Grail,* chap. 2, 60.
25. Wilson and Prentis, *Power of the Magdalene,* 124.

To the Burned, the Slaughtered,
The Tortured and the Persecuted.
For all who have suffered and all who have died
At the hands of Darkness and its Inquisition,
But never has the Darkness overcome them.
After 700 years the Vine turns green again.[26]

---

26. *GBC*, iii.

Ascension in the Twentieth and Twenty-First Centuries

Figure 19.0. William Blake, *Jacob's Dream*, 1805,
The British Museum, London.

19

ASCENSION:
THE SCIENTIFIC EXPLANATION
OF QUANTUM PHYSICS

Mary Magdalene was not the only one who made an ascension. The religious and mystical history of humanity is full of examples of other people making an ascension or a heavenly ascent. Many ascension experiences were recorded and handed down throughout ancient cultures, including Ancient Egypt, Mesopotamia, Greece, India, Tibet, and China. In later times there were Hebrew prophets who ascended as well as Buddha, Yeshua, his mother Mary, Mary Magdalene, Muhammed, Saint Francis, and several other Christian and non-Christian mystics. Ascension occurred internationally and transculturally and was therefore a phenomenon in itself and yet is universal.[1] Ascension is something fully human. To become Human with a capital letter means to become Light. Making yourself truly human and putting on perfect humanity, as stated in Miryam's Gospel, means again becoming that being of light that you once were.[2]

In the past, ascension was considered solely a religious phenomenon. In the enlightened West it was seen as something for floaters and mystics, for people who fell prey to religious madness, something for people who belonged in an asylum. Those days are now definitely over. In the

---

1. Greene, "The Celestial Ascent," ii and 2.
2. *GBC* [41:3], 76 and [43:6], 81; *MMU*, 449, 457.

twentieth century there are new scientific explanations about ascension that arise from new insights about quantum physics.[3]

The majority of Western people still reason from philosophical materialism. This assumes, among other things, that the universe consists purely of chance combinations of material particles. Man would consist of only a body; soul and spirit would be fables from the distant past that no one can prove. That now seems to be changing. Several researchers who serve the new science bridge the gap between science and spirituality. From different disciplines they build bridges between the visible and the invisible, between the physical and the nonphysical or metaphysical, between spirit and matter. They see consciousness as an expression of spirit in matter.

Bridge Builders Between Science and Spirituality

It is not the intention to give a total overview of all these bridge builders. Hereafter, only a few people are presented who have developed insights about subtle energy and ascension.

Russian scientists Valerie and Semyon Kirlian invented a new photography method in which they succeed in photographing certain subtle energy fields, including those around plants and animals. They photographed the aura and how it responded to emotional and psychological states.[4]

Valerie Hunt developed electromyography and built an instrument that measures up to 750,000 Hz. Working with the electromyograph, she found that the physical body emits rays of colored light precisely in those places where ancient cultures perceived colored wheels with their inner eye. In other words, she scientifically proved the existence of the chakras. It was established that there are seven base layers that corre-

---

3. Greene, "The Celestial Ascent," 70: "The language of contemporary ascents is certainly new in that it incorporates modern psychological terminology and scientific theories adopted from such diverse sources as quantum physics and the older models of Mesmer and Swedenborg."

4. Dale, *The Subtle Body*, chap. 27; Rennison, *Afstemmen van op de Kosmos*, 110; Ash, *Superenergy and the Quantum Vortex*, 87

spond to seven colors and tones. The colors correspond to various hertz frequencies.[5]

The HeartMath Institute in California researches the heart. It found that the heart produces a large energy body around itself from which a small light emanates (Figs. 19.1a–b). That light generates a field around the heart, but only when it is open and happy.[6] Positive emotions create coherence waves, which form a coherent pattern. Then, other organs begin vibrating with that coherence wave pattern. When humans shift attention from the head to the heart and the mental chatter of the head comes to rest, we become more intuitive.

Anger creates incoherent and irregular waves in the heart, blocking the brain. The communication between the heart and head and the rest of the body is disrupted. A chain reaction takes place: the blood vessels constrict, the blood pressure rises, the immune system weakens, the brain produces high beta waves, and diseases can arise.

Figures 19.1a and 19.1b. The heart chakra with the two triangles of feminine and masculine energy in perfect balance. The HeartMath Institute in California recently scientifically established that a small light emanates from the heart chakra. When the heart is open and happy, it generates light.

---

5. Dale, *The Subtle Body*, chap. 27; de Vries, *The Whole Elephant Revealed*, 111–12, 153.
6. See HeartMath Institute website; Rennison, *Afstemmen van op de Kosmos*, 117, 129–33; de Vries, *The Whole Elephant Revealed*, 155n32.

David Hawkins, in his book *Power Versus Force,* developed a scale on which emotions were represented, ranging from a strong imbalance to a complete balance.[7] The energy fields of consciousness that were represented by him scale from 20 to 1000. The area around 200 marks an important transition to truth, integrity, and courage. It goes from shame at 20, through apathy at 50, fear at 100, anger at 150, pride at 175, to courage at 200. Neutrality is at 250, acceptance at 350, love at 500, joy at 540, peace at 600, and enlightenment at 700–1000. The scale is logarithmic, so an increase of a few units means a large increase in strength. When you develop spiritually, you increase the extent to which you influence other people around you.

In 2001, Lynne McTaggart included a substantial chapter in her book *The Field: The Quest for the Secret Power of the Universe* entitled "We Are Beings of Light." The main person discussed in this chapter was the biophysicist Fritz-Albert Popp, who was affiliated with the International Institute for Biophysics in Neuss, Germany.[8] In 1976, using highly sensitive equipment, he discovered that living cells emit photons or biophotons and that every living cell emits a pulse of light as many as 100,000 times per second; they are an aspect of the human energy field. On the one hand, biophotons form a perfect communication system to transfer information to the many cells within an organism. On the other hand, they also function as a communication system between different organisms. The photons emitted by healthy living organisms are coherent and in phase with each other.[9] Health appears to be a state of coherence and disease a state in which the coherence is disturbed and the wave motions are out of phase. Popp said, "Today we know that man is essentially a being of light. And the modern science of photobiology is currently proving this. We now know that light can start or stop cascading reactions in cells and that genetic damage to cells by weak light rays can be virtually repaired within hours."[10]

---

7. Hawkins, *Power Versus Force*; de Vries, *The Whole Elephant Revealed*, 121–22.
8. McTaggart, *The Field*, chap. 3,
9. de Vries, *The Whole Elephant Revealed*, 142, 153.
10. See "Dr. Fritz Albert Popp," Biontology Arizona website.

Physicist David Bohm and the Three Realities

The British physicist David Bohm (1917–1992) deserves special attention. In his 1980 book *Wholeness and the Implicate Order* he started from the hologram, an organized field in which each layer closer to the center contains a higher form of information. The special thing was that each part (fractal) contains the information of the whole. The whole can be found in the parts. The patterns of the macrocosm and microcosm are the same. Bohm knew the holographic reality of unbroken wholeness; it was, however, divided into three main structures made up of three layers. From a Source of Unity that at the same time contains an infinite reservoir of energy, the Web of Life flows forth in three realities.

1. The unmanifested reality: Bohm spoke of formless and invisible reality, a hidden order. He called this the super-implicate order, super in the sense of "above," so the order above the implicate. In Miryam's Tree of Life this was the area above the eighth level and in other texts above the thirteenth sphere. You could also call it the nonphysical world of Spirit. Alariel previously spoke of The Unmanifested.[11]

2. The implicate reality: this is also hidden and not visible and is, as it were, folded into the unmanifested order. In Miryam's Tree of Life this is the area from level 5. You could call it the subtle world of the Soul. Alariel spoke of the Greater Reality or Middle World.

3. The explicate reality: the unfolded or manifested world of space and time. In Miryam's Tree of Life, this would be the area at the roots of the tree, the physical world of the third and fourth levels. Alariel spoke of the Lesser or Limited Reality. You could call it the dense physical world of the body.

---

11. Wilson and Prentis, *Beyond Limitations*, 99.

David Bohm considered the factor "consciousness" as the primary factor in the three realities.[12] Everything flowed from consciousness or information. From consciousness comes energy, and energy is a manifestation of consciousness. Energy expresses itself in matter at a lower vibrational level. David Bohm defined matter as condensed consciousness.[13] From the source of Oneness and the Web of Life, consciousness flows forth, manifesting as energy in a lower vibration, energy manifesting as matter in an even lower vibration.[14]

The explicit order is septenary. At lectures I like to use a series of seven *matryoshkas* that I once brought home with me from Russia. I'd say, "Here you see seven matryoshkas. The smallest minimother is encompassed by much larger ones. Thus, the limited reality is enveloped by the greater reality, and it is clothed again with the Unmanifested. Now we go the other way around. Imagine a funnel: the Unmanifest expresses itself in ever less spacious and more limited forms until the physical is reached in the smallest and seventh mother."

I draw the following comparison: "The physical appears to be made up of seven states. Let's assume water: in its most solid form or state it has solidified into ice. When the environment becomes warmer, ice turns into liquid water. If that environment is heated even more, water turns into steam. These are the first three grossest forms that matter can take: solid-liquid-gas. Not everyone knows that matter can take on four more states that become finer and more subtle with ever smaller particles. After solid-liquid-gas you reach the ethereal realm. It consists of two finer forms of matter: 4. Etheric (plasma or highly heated gas) and 5. Subetheric (more highly heated plasma). You reach even finer structures in the atomic and subatomic realm. That gives 6. Atomic

---

12. Goswami, *The Self-Aware Universe*, 108 and 270 concludes that consciousness is the ground of all things; Ash, *Awaken*, 192–193 also refers to Max Planck who regarded consciousness as fundamental; Ash, *Superenergy and the Quantum Vortex*, 52

13. Friedman, *Bridging Science and Spirit*, 317; de Vries, *The Whole Elephant Revealed*, 101n12.

14. de Vries, *The Whole Elephant Revealed*, 66, 101, 173–74.

Figure 19.2. The Seven Matryoshkas.

and 7. Subatomic[15]" (fig. 19.3). This is the septenary division of the seven states in which matter can manifest. A human being has a dense body and an etheric double or form body. The smallest particle is the subatomic particle. How does this take on a higher and faster form of vibration? How can it take off? How can you rise from the bottom of the funnel from the limited to the wider or greater reality? The key lies in the wave that becomes a vortex of energy.

The Physicist David Ash Developed the Physics of Ascension

There is a remarkable British scientist who has devoted himself to vortex energy.[16] From there he developed the physics of ascension.[17] His name is David Ash.

---

15. Wyatt, *The Cycles of Eternity*, 21 with images.
16. Ash, *Superenergy and the Quantum Vortex*, 5–11.
17. Ash, *Awaken*, 76.

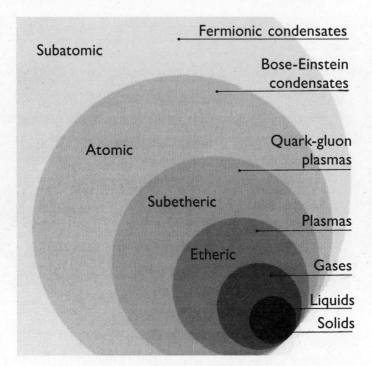

Figure 19.3. The lowest or physical-etheric layer has
seven states of matter:

1. Solids, physical   5. Quark-gluon plasmas, subetheric
2. Liquids, physical   6. Bose-Einstein condensates, atomic
3. Gases, physical   7. Fermionic condensates, subatomic
4. Plasmas, etheric

From Tim Wyatt, *Cycles of Eternity: An Overview of the Ageless Wisdom*, 23.

Physicists have discovered that atoms are made up of vortices, min-
iature vortices, or whirlwinds of energy that constantly spin and vibrate.
A collection of such infinite small energy vortices together forms an
atom. Matter is described as "vibrational waves revolving in circles at a
place."[18] David Ash's story revolves around the wave and the circle. It
is a wave that starts spinning in a circle and spirals upward in a mael-

18. de Vries, *The Whole Elephant Revealed*, 100.

strom or whirlwind: a vortex. The fact that waves of particles spin faster and faster in a vortex is the basis for David Ash's physics of ascension. Everyone knows that when the pilot of a helicopter wants to take off, he pushes a button to speed up the blades or propellers. Ash compares the vortex to a spherical structure and preferably to a rolled-up ball of wool, although the comparison is not entirely true. Here, the line spirals out from one point, with the poles continuously changing, which creates a spherical shape.[19] An enormous amount of energy can be stored in this.

The Life of David Ash

David Andrew Ash was born in 1948 in Kent, England. He was the son of a doctor who also paranormally healed his patients and who in his time researched the relationship between cancer and radioactivity.[20] When David was four years old, he decided to prove God through science. His father introduced him at this young age to atomic physics and explained the relationship between healing and nutrition. In 1965, at the age of sixteen, in the attic of a private library that was filled to the brim with antiquarian books, Ash found an old book from 1905 (before Albert Einstein had formulated his theory of relativity) about the philosophy of yoga and borrowed it.[21] There he read that energy, known in India as prana, is present in matter in the form of whirlpools or vortices. That which comes across as mass is actually a form of energy in a certain vibration and rotation or spin. Ash already understood at the age of sixteen that subatomic particles were vortices of energy and light and not solid matter. Energy forms mass by spinning. Energy or light forms matter by spinning. This book was a real eye-opener for Ash.

Ash describes himself as an ascension physicist, singer and songwriter, international speaker, and nutritionist. He has dedicated his life

19. Ash, *Awaken*, 340; Ash, *Superenergy and the Quantum Vortex*, 199.
20. Ash, *Superenergy and the Quantum Vortex*, 83
21. Yogi Ramacharaka, *Advanced Course in Yogi Philosophy and Oriental Occultism*; Ash, *Awaken*, 9, 236, 281, 284, 327, 379; later Ash discovers that the book was written by an American named William Atkinson (1862–1932).

to developing a new vortex physics to connect and reconcile science and spirituality. Through his discoveries, he realized the possible existence of higher dimensions of reality beyond the speed of light. It involves a process where bodies of matter rise from physical space and time into worlds of living light, of which our visible world is only a limited part. In this sense, Ash discovered the "physics of ascension."[22] The possibility of ascension first occurred to Ash in the early 1970s. He deduced it from his vortex theory. In 1975, he presented his vortex theory at the Royal Institute of Great Britain, and as a physicist, he predicted that the physical is moving toward spiritual levels. He published the idea of the acceleration of energy as an alternative to death in *The Tower of Truth* in 1977. He used this knowledge to explain the ascension of Jesus and other masters in *Science of the Gods* in 1990.

Ash and Einstein

In classical antiquity, Greek philosopher Democritus (460–370 BCE) stated that the smallest element was the atom. In his view, atoms were indivisible, eternal, and unchanging. In addition to being considered the founder of atomism, he was also considered the father of materialism.[23] He learned that everything in the universe was created by chance encounters of indestructible atoms, which, in his eyes, the tiniest particles of matter acted like mini billiard balls. Later, Einstein and others discovered that atoms are divisible and consist of many subatomic particles with a nucleus and with protons, electrons, and neutrons. Energy whirls around and clumps together into matter. They discovered that the basis of the material world is not matter, but energy, but despite this, materialism remains popular.

In experiments with subatomic particles, quantum physicists have demonstrated that certain fields appear to contain a kind of intelligence or information that plays an active role in the energy–matter relation-

---

22. Ash, *Awaken*, 76; Ash, *Supernergy and the Quantum Vortex*, 5, 7. 198.
23. Ash, *Awaken*, 186, 192, 280; Ash, *Superenergy and the Quantum Vortex*, 8, 28.

ship. They believe that the factor of information or consciousness must be involved in the interaction between energy and matter.[24] In his theory of relativity, Albert Einstein emphasized the importance of the speed of light in our physical world. He said that space and time are not absolute and that mass, space, and time are relative and subject to the speed of light.[25] Light has been measured to travel at 299,792,458 meters per second.[26] This speed of light was, according to Einstein, the only absolute and universal constant. It is therefore important—based on Einstein's theory of relativity—to establish the following: there is no absolute universal space or time.

Einstein turned classical scientific materialism on its head with his formula $E = mc^2$. Energy (E) at the speed of light (c) is the basis of mass (m). Einstein revealed that atoms in matter were made up of tiny particles of energy and that these were not particles of material substance but were made up of energy or light. He discovered the nonmaterial, nonsubstantial nature of matter.[27] He realized the consequence of his theory: energy in the form of small particles, moving at the speed of light, forms the basis of everything, including mass, space, and time. To summarize: mass, space, and time depend on light and the speed of light. Energy at the speed of light is the basis of mass. Energy forms mass, time and space.

Einstein, according to Ash, was one of the first in the West to recognize the error of philosophical materialism.[28] His theory of relativity put an end to classical scientific materialism, which assumed that the basis of the universe was material. Some people preceded Einstein in discovering that materialism is a myth.[29] Those were the Indian yogis.

The yogis clairvoyantly observed that particles were spinning, after which fundamental particles of matter were formed. They realized that

---

24. de Vries, *The Whole Elephant Revealed*, 173.

25. Ash, *Awaken*, 102; Ash, *Superenergy and the Quantum Vortex*, 6, 11.

26. Ash, *Awaken*, 77.

27. Ash, *Awaken*, 189; Ash, *Superenergy and the Quantum Vortex*, 11.

28. Ash, *Awaken*, 198; Ash, *Superenergy and the Quantum Vortex*, 6 .

29. Ash, *Awaken*, 199.

these vortices of light were at the root of the material forms.[30] In around 700 BCE, Pakistan had a university at Takshashila where 68 disciplines were offered to 10,500 students from India, Arabia, Babylonia, Greece, Syria, and China (see an Asian Interlude). Mathematics was well developed here. For example, a trillion is ten to the power of twelve. Modern mathematicians go up to ten to the power of thirty. But there is an Indian sutra from 100 BCE in which an awesome number is mentioned with great precision from 10 to the power of 140, as far as they could go. Pythagoras studied in India among other places. According to Ash, many of the inventions attributed to the Greeks originated in India. Ash stated that a few centuries ago, Western philosophical materialism explained away mysticism. Now the mysticism of India explains away philosophical materialism.[31]

Modern physics focuses on the study of elementary particles in subatomic physics. The subatomic particle would in turn consist of many particles. Physicists continue to search for ever smaller physical particles such as the quark, the lepton, and the Higgs particle. In David Ash's view, they do everything they can to prove that the very tiniest particle is material. Since 1960, so many new particles have been found that today we speak of the "particle zoo."[32] Scientists have searched frantically, further and further, for the material particles that form the bottom of matter, and pay large sums for research.

According to David Ash, Albert Einstein would have regretted this development because he did not like the uncertainty principle as developed by quantum physicists at the Copenhagen Institute of Physics: Werner Heisenberg, Wolfgang Pauli, Max Born, and Paul Dirac. Why not? Because it was precisely because of the built-in uncertainty factor that it was not possible to prove the theory wrong.[33] Einstein disliked the idea that the quantum world was essentially unknowable and felt that quantum physics was going in the wrong direction. In his opinion,

---

30. Ash, *Awaken*, 273.
31. Ash, *Awaken*, 282–83, 289.
32. Ash, *Awaken*, 247–248, 249, 298; Ash, *Superenergy and the Quantum Vortex*, 9.
33. Ash, *Awaken*, 207, 247, 265, 267, 279.

the uncertainty principle was being misused to further develop quantum physics,[34] giving rise to speculative theories.[35]

According to Ash, everything is explained by the fact that small particles in active interaction interact with each other, but an underlying field of energy and consciousness is denied, despite the yogis seeing vortices of light and energy or prana from which form arises. When light spins, it forms matter. Ash said, "How can you ride the wave of the uncertainty principle when you don't acknowledge the ocean?"

According to Ash, an energy wave is drawn into the spiraling path of the vortex like a car following the curve in the road. Quantum physicists have tried to explain the universe by assuming only the wave, which manifests itself in a line. But this, according to Ash, is only half the story: the vortex of energy is the other half. That vortex starts from the circle, and when that circle moves up it creates a spiral. The linear wave ends in a circle that develops into a spiral. At the beginning of the twentieth century, the vortex theory, which was already developed in England in Victorian times, became popular. However, it was applied on the atomic level, so the theory was released because it didn't work on an atomic level. Ash applied the older vortex theory at the subatomic level, and then it worked.[36] The protons, electrons, and neutrons in the atom create the positive and negative tension fields in which infinite vortices of energy are created.[37]

Ash called the linear wave masculine and the spiraling circle feminine. They need each other to regenerate tension; you can compare

---

34. Ash, *Awaken*, 247: The neutron was discovered in Cambridge in 1932; it rehabilitated Einstein and proved the uncertainty principle false; 254: A neutron can be composed of an electron or a proton and of opposite charges, which makes the neutron unstable in contrast to the proton and the electron; however, modern students are taught that the proton is composed of two up quarks and a down quark while the neutron is said to be composed of two down quarks and an up quark. According to Ash, everything revolves around the neutron that changes shape into a proton and electron. The search for smaller particles can stop.

35. Ash, *Awaken*, 269.

36. Ash, *Awaken*, 77, 290, 328.

37. Ash, *Awaken*, 309.

it with the sperm cell and the egg cell—both work together.[38] The wave and the vortex make a marriage of male and female forms of energy. When you reintroduce the vortex theory, according to Ash, you reintroduce the feminine principle and give her a rightful place. Ascension is not only about vibration and frequency, but it is also about couples of waves and vortices that can only reach a higher vibration together.[39]

In addition, there is another principle at play with the particles that manifest themselves in waves and vortices at the subatomic level: the principle of nonlocality. There is an invisible cause beyond space and time that allows particles to instantly influence each other regardless of geographic distance. Particles form compounds that exist in a reality beyond space and time.[40] According to David Ash, his vortex theory made sense of the immaterial part of reality. The particles are entangled and connected to each other (interdependence).[41] They attract and repel. Everything stands or falls with mutual interaction. Ash called the universal law of love the driving force behind everything.[42]

Ash didn't go along with Einstein on everything. Einstein stated that the speed of light was the sole constant.[43] This, according to Ash, may be true for our physical 3D world, but not for the higher worlds, which could vibrate at several higher speeds of light. Think of the blades of the helicopter: when the propellers start spinning faster, the helicopter takes off. The 5D world would have light speeds that are much higher than the speed of light. Ash named a number three times the speed of light or more. This would not contradict Einstein's claim that the laws of physics are determined by the speed of light in this 3D world.

38. Ash, *Awaken*, 315–17.
39. Ash, *Awaken*, 315.
40. de Vries, *The Whole Elephant Revealed*, 223–24.
41. Ash, *Awaken*, 104.
42. Ash, *Awaken*, 103–04.
43. Ash, *Awaken*, 77.

The Three Planes

Like David Bohm, David Ash distinguished three levels.

1. The physical or 3D level of the physical body.
2. The hyperphysical or 4D level of the soul: this is the level we enter after death. It forms an intermediate or interim stage between physical incarnations.
3. The superphysical or 5D level and higher of the mind.[44]

All forms that vibrate slower than the speed of light belong to the physical universe, according to Ash. The second reality would represent the vibration of twice the speed of light. From 5D it would be three times and more. When the energy accelerates, higher frequency levels are created.[45] Ash compared it to the speeds of a bicycle, a car, and an airplane. You cannot overtake a car on a bicycle, nor can a car overtake an airplane. Slow speeds are part of higher speeds but cannot keep up with them.[46]

Ash used the matchbox analogy to illustrate the relationship between the planes of super-energy and the plane of energy we live on. He depicted the physical world as a matchbox that is in a room. The matches in the box represent the people on Earth; they are only aware of the world within the box. They are unaware of the room with creatures outside the matchbox. In turn, the creatures in the room are oblivious to the fact that the room is part of a house. You can't put a room in a matchbox or a house in a room. Different laws apply here because the speed of light differs.[47]

---

44. Ash, *Awaken*, 51. This is a different format than the one where we previously encountered Miryam's Tree of Life. There 3D and 4D were counted among the roots of the tree; the middle area of the soul started from 5D and went up to 8D, and above that the world of spirit started.

45. Ash, *Awaken*, 78–81.

46. Ash, *Awaken*, 81.

47. Ash, *Awaken*, 82; Ash, *Superenergy and the Quantum Vortex*, 17.

According to Ash, the universe is like a staircase: each step is a separate space-time center with different energy velocities. It consists of an ascending series of quantum realities based on different speeds of energy. The 3D reality has three spatial dimensions and the factor of time that extends from the past through the present to the future. In 4D, space and time seem to merge into unity.[48] Between the fourth and fifth dimensions there is a barrier. Ascended beings live in the so-called fifth dimension.

God as Consciousness

Ash wrote "Long before Richard Dawkins wrote *The God Delusion*, belief in God fell out of favour. Today, many think that people like me who still believe in God are deceived. But unfortunately many people are out of touch with physics in their thinking." Then Ash quotes Cambridge physicist Sir James Jeans from *The Mysterious Universe* who said: "Today there is wide measure of agreement that the stream of knowledge is heading toward a non-mechanical reality; the universe begins to look more like a great thought than a great machine. Mind no longer appears as an accidental intruder into the realm of matter; we are beginning to suspect that we ought rather to hail it as the creator and governor of the realm of matter.[49] There cannot be a thought without a thinker so if to enlightened physicists the universe appears to be a great thought, then it would not be unreasonable to assume there is a great thinker behind the great thought and call that great thinker of the universe, God."[50]

Ash continued: "Something that is not accepted in science is using a term without defining it. The word God is acceptable in science if it is given a scientific definition. God could be defined as Universal Consciousness. In May 2020, The *New Scientist* announced that scientists have discovered that Consciousness is a Universal principle and that

48. Cinamar, *Forgotten Genesis*, 57.
49. Ash, *Awaken*, 188; Ash, *Superenergy and the Quantum Vortex*, 7, 52.
50. Ash, *Awaken*, 188.

the idea that Universal Consciousness is the ground of all being, is the starting point of everything. This has been accepted by leading quantum physicists for over a century. To quote Max Planck, the founder of quantum theory: 'I consider consciousness as fundamental. I consider matter as a derivative of consciousness. We cannot get behind consciousness.'"[51]

The Physics of Ascension

Looking back at his life's work, Ash explained that he worked for more than half a century on a vortex approach to physics that counters materialism. Vortex physics can include spiritual and psychic phenomena and explain many things that materialists have tried to explain away. This is because it opens up the possibility that there are worlds we are not yet aware of, worlds that are invisible and elusive to us because they are formed from energy that moves faster than light.

In theoretical physics, the speed of light is considered the fastest possible speed. It supposedly would be proven that "material" particles cannot move faster than this speed. But in some parts of the universe, energy may be moving faster than light. It flows in waves and swirls around in vortices, to form worlds based on these higher speeds of energy, which I call super-energy. Super-energy at different speeds could be the basis of the higher dimensions. Higher rates of energy represent higher frequencies of vibration. The idea of super-energy dimensions is just a prediction but predictions are the foundation of science. They drive experiments, direct discovery, and feed investigative minds with new and exciting possibilities. The physics of ascension is based on this prediction; and the ascension would be the living proof.

Ash went on to explain that:

The third, fourth and fifth dimensions, represented by body, soul and spirit, can be depicted as a set of concentric spheres.[52] This

---

51. Ash, *Awaken*, 191–92.
52. Ash, *The Door of Light*, 24.

model of the physical world contained by spiritual worlds came from ancient times. Pythagoras described it as the Harmony of the Spheres. This arrangement of the dimensions can be appreciated today from something Einstein realized about energy.

Each dimension has its own space-time continuum set up by its

Figure 19.4. The world of light shows itself in different rings. Engraving by Gustave Doré (1832–1886) of *The Divine Comedy*, *Paradiso*, Canto 31 by Dante Alighieri.

own level of energy or super-energy. The dimensions are in simultaneous existence as they are not separated by space or time but by the speed of their energy—or frequency of vibration. Hermes called the dimensions *planes* which is where the word 'planet' comes from. Hermes also coined the word 'human' from the Egyptian word *Hu* for god or divine. He described us as *God-man* or Divine-man beings because we are multidimensional, that is we are simultaneously conscious in body, soul and spirit. From my physics I am confident we are alive in a third dimensional physical body, a fourth dimensional soul body and a fifth dimensional spirit or angel body. This is possible because our bodies in the different dimensions exist simultaneously, i.e. they can overlay each other as there is no space-time separation between them. The overlapping bodies then interact with each other through resonance. Our consciousness resides in the spirit body which resonates with the soul body. The soul body, which resides with the physical body, resonates with it through the DNA molecule. I call this process DNA Resonance, which is explained in *Awaken*.[53]

The Physics of Ascension Explains Paranormal Phenomena

The next step Ash makes is constructing a bridge between the physical and nonphysical levels. According to Ash, with the science of ascension, a great many paranormal and miraculous phenomena can be explained. Major world religions have examples of ascension: Zarathustra in the Persian tradition; Elijah in the Hebrew tradition; Muhammad on Mount Moriah in Islam; and Jesus on the nearby Mount of Olives in Christianity. These prophets would not die and have conquered death, nothing else is explained. Ascension as the alternative to death is part of several religious traditions. But according to Ash, metaphysical concepts are clarified by a new approach to

---

53. Ash, *Awaken*, 218-23.

quantum physics; because of this, the mystical loses the mysterious and the physical ascension is presented as a logical and scientifically based phenomenon.

From 1968 to 1970, an apparition of Mary, the mother of Yeshua, hovered over the dome of a Coptic church in Zeitoun, a suburb of Cairo, Egypt. Night after night, for well over a year, the outline of a Lady with a halo appeared in light emitting ectoplasm which poured from her luminous human figure, and flowed over the roof and dome of the church, lighting up the night sky with its brilliance.[54]

Millions witnessed this because the image was broadcast on television worldwide. Ash saw it in England on the early BBC evening news and explained it this way:

Every object in our world is made of trillions of minute particles, once called atoms, now called subatomic particles. In the vortex physics these appear to be whirlpools of light or vortices of energy where spin gives rise to an illusion of materiality. This is the cause of The Material Delusion. I explain miracles, like the apparitions of Zeitoun, in terms of altering the speed of the energy in the vortices. Mary, in her fifth dimensional ascended body, could have used a technology that I call super-energy resonance to depress the speed of energy in every subatomic vortex of super-energy in her body to the speed of physical energy. That would have enabled her to condense into the third-dimensional world at Zeitoun.

When the door of light appears to us we will be looking into a super-energy resonance beam. The super-energy in the beam would be at the speed of fifth dimensional light. As we step into the beam, the energy in every subatomic vortex in our body would resonate and then accelerate to the speed of super-energy in the beam and so become super-energy. To enter the fifth dimension we will not go anywhere, as such. It is more we will quicken; that is the energy in our body will accelerate to become super-energy. After our spirit is

---

54. Ash, *Awaken*, 152; Ash, *Superenergy and the Quantum Vortex*, 112.

awakened to the higher dimension, it will be transformed . . . and the spiritual energy in our body will rejuvenate it so that we become eternally youthful. We will also be able to use the resonance process to travel by evaporating out of and condensing back into physical space.[55]

55. Ash, *The Door of Light*, 22–28.

20
THE CONSCIOUSNESS
REVOLUTION

During the Enlightenment in the eighteenth century—also called the Age of Reason—ascension or heavenly ascent faded into the background because it no longer fit into the materialistic and rationalistic worldview of the Enlightenment. Ascension was grouped together with shamanism under the heading of archaic superstition, and because of this ascension was neglected for a while by science.

At the end of the nineteenth century, theosophy and anthroposophy brought the ascension traditions back into Western collective memory. Within the New Age movement from 1960 onward, an explosion of interest in Eastern spirituality and inner growth followed, which has had an impact on the spiritual movements of the twenty-first century. You could describe the period from 1875 on as a Revolution of Consciousness.

The Founding Mothers and Fathers after 1875: Modern Theosophy and Ascension

After 1875, with a renewed interest in mystical phenomena, people became interested again in the process of ascension. At the beginning of this development, which was also labeled an acceleration in consciousness, was Mrs. Helena Petrovna Blavatsky (1831–1891). She published her first book, *Isis Unveiled,* in 1877, and it consists of two parts that together total an impressive 1,648 pages. Later she wrote to her cousin that while writing she felt overshadowed by the Goddess Isis.[1]

In 1888, *The Secret Doctrine* followed, a book of 703 pages, in which she wrote that she was inspired by the masters. In doing so, as a courageous pioneer she laid the foundation for the further expansion of consciousness in the twentieth century, at a time when Western male intellectuals were under the spell of philosophical materialism and most women were considered too simple to be educated. Mrs. Blavatsky popularized Eastern ideas about reincarnation, karma, and the existence of master souls; she introduced these concepts into Western culture. It is true that an esoteric undercurrent carried these ideas to the West from antiquity to the nineteenth century, but it had receded into the background within mainstream thinking. Blavatsky—often maligned by gossip and backbiting—reintroduced these ideas to the West and permanently placed them on the intellectual and spiritual agenda.[2]

Mrs. Blavatsky obtained the information for her books and many articles through nonphysical ways and channels. Among other things, she was connected to ascended masters from higher dimensions and knew herself to be guided by them.

Alice Bailey, a theosophist of the second generation of theosophists, said she was inspired by the ascended master Koot Hoomi. At the time when Mrs. Blavatsky and Alice Bailey communicated with the masters—in fact they channeled the information—this form of information transmission from a higher dimension still seemed to be reserved for a few, but that changed in the twentieth and twenty-first centuries.

The philosophy of Helena Petrovna Blavatsky started from the heavenly ascent of the soul. In her visions, the soul, after death, moved toward the eternal source. The temporary bodies or vehicles the soul thereby discarded on its way up were related to the different levels of consciousness; they were reflected by the seven stages the adept must pass through to spiritual enlightenment.[3] When the soul descended

---

1. Ingmar de Boer, *De Stem van de Stilte: De Wereld achter het werk*, Amsterdam: Theosofische Vereniging, 2013, 8–10 and 11–12.
2. Wilson and Prentis, *Beyond Limitations*, 73–74.
3. Greene, "The Celestial Ascent," 65n59.

from that source, it clothed itself in a causal, a mental, and an astral body. These bodies were a kind of vehicle in which the soul descended into the condensed physical body. That body also had an etheric double with an etheric field directly surrounding the body.[4]

All of these levels manifest on both micro and macro levels; they are also areas in the cosmos on a macro level. They do not exist in the physical universe but outside it in the subtle and immaterial worlds. The idea that the soul clothed in garments or vehicles ascends through successive levels has, according to the Dutch professor Wouter Hanegraaff—a specialist in scientific New Age studies—penetrated from theosophy into many contemporary spiritual movements.[5]

In fact Madame Blavatsky brought universal esoteric knowledge back to the West via the East. Others followed Blavatsky, such as Rudolf Steiner and Alice Bailey; they consolidated and expanded Blavatsky's legacy in their own way. Without these contributions, many healing techniques and therapies later incorporated into the New Age movement would never have occurred (again according to Wouter Hanegraaff).[6]

Carl Gustav Jung and Ascension

C. G. Jung (1875–1961) wondered whether the celestial ascent was a macro or micro cosmic journey. Was it a journey in the macro cosmos through the planets, or a journey on a micro cosmic level through the chakras, a journey within? Jung formulated a compromise. The macrocosmic experiences were both external and internal, both real and imaginary.[7] Jung—like philosophers in antiquity and the Middle Ages—made a connection with alchemy. The seven planetary spheres were associated with the seven phases of the alchemical process. Jung

4. Greene, "The Celestial Ascent," 66n361.
5. Hanegraaff, *New Age Religion and Western Culture* 95, 260, 448, 518.
6. Hanegraaff, *Dictionary of Gnosis and Western Esotericsm*, "New Age Movement" entry, 855–861, entry A: Healing, 856–57; Hanegraaff, *New Age Religion*, chap. 2 "Healing and Personal Growth," 42–61; also see "holistic health," 42, 53, 242.
7. Greene, "The Celestial Ascent," 66n362.

was concerned with the individuation process, the individual becoming healed and whole; it was about the soul transforming lead into alchemical gold. It was about balancing the good and bad qualities to the maximum of consciousness and with the maximum use of free will. The integration of the aspects led to individuation in which the separated parts of the shadow or unconscious connected with consciousness and become integrated and whole.[8] Jung psychologized the heavenly ascent. He saw this as an archetypal process of individual development, a process that took place on both a psychological and a spiritual level. From an astrological and alchemical context, the heavenly ascent now develops into a psychological and spiritual phenomenon.[9] Jung formed a crucial link with the later New Age movement that manifested itself starting in the 1960s and 1970s.[10]

Ascension in the Twentieth Century: The New Age Movement

From 1960 and 1970, the contours of a new impulse became visible; it was a second acceleration in consciousness. The new movement was given the legendary name *New Age* by theosophist Alice Bailey.[11]

With the New Age movement, churches were being pushed out of their monopoly on pastoral care. There was a transition from religion to spirituality; the new spirituality was individual and could not be captured in a dogma or hierarchical institution. There was no New Age creed or a New Age bible. New Age formed a web of connections, a global network, each following their own chosen path to the light. It was an autonomous spirituality, which was very different from organized religions. Many small streams and rivers come together in a great stream of wisdom from many traditions. This was the reality of the

---

8. Greene, "The Celestial Ascent," 76; Hanegraaff, *Dictionary of Gnosis and Western Esotericsm*, entries for "Jung," 648–53 and "Jungism," 651–55.
9. Hanegraaff, *New Age Religion and Western Culture*, 496.
10. Hanegraaff, *New Age Religion and Western Culture*, 513.
11. Hanegraaff, *New Age Religion and Western Culture*, 95.

second half of the twentieth century.[12] In the twenty-first century it swelled into many new spiritual movements.

The Bridge Between Science and Mysticism

In the twentieth century, quantum physics took off. In the previous chapter, physicist David Ash was mentioned: he had focused on a scientific explanation for ascension. For him, ascension was a natural process that can be explained by the physics of ascension as previously discovered and taught by mystics in India. Ash thus formed a bridge between both fields: physics and spirituality.

It was only after quantum physicist David Ash had developed the physics of ascension that he came into contact with the channeled literature about ascension. In the introduction to his book *Awaken*, Ash wrote that he had felt all his life that a great change was coming for humanity, a massive spiritual transition into a New Age of Light.[13] He lists ascension experiences prior to the New Age movement[14] and describes (and gives examples of) how the information about ascension emerged in the 1930s.[15]

In the following decades—especially after 1960 within the New Age movement—more and more information about ascension followed.

- Jane Roberts channeled the interdimensional teacher Seth, who began communicating with a completely unprepared Jane beginning in 1963. A first publication followed in 1974.
- Beginning in 1965, Helen Schucman channeled the 1,188-page book titled *A Course in Miracles*, published in 1975.

---

12. Wilson and Prentis, *Beyond Limitations*, 75.
13. Ash, *Awaken*, 9.
14. Ash, *Awaken*, 9.
15. In the 1930s Godfrey Ray King channeled the ascended masters. In the 1950s Sister Thedra received information about ascension. In the sixties and seventies there were the Mark Age channelings.

- In 1977 James Hurtak published *The Book of Knowledge: The Keys of Enoch.*
- In the 1980s there were publications about ascension by the authors Tuella and Mark Prophet, founder of The Summit Lighthouse in the US.
- In 1979 Ken Carey received *The Starseed Transmissions.* It mentioned the year 2011 as the date for the second coming of Christ. Humanity would awaken to its innate divinity and millions of people would come to their innate Christhood without a single individual Christ returning to Earth. It would be about the awakening of humanity in Spirit. It would involve an ascent of humanity into Spirit and a descent from Spirit into man.

Eric Klein's Ascension Tapes

Yet the physicist David Ash only started to really understand the phenomenon of ascension in a spiritual sense on November 11, 1991. It is after his arrival in Australia that David Ash became acquainted with Eric Klein's five cassette recordings, the so-called Ascension Tapes. The tapes described the possibility of an individual and collective ascension that could precede large-scale Earth changes. In the tape recordings, people were said to be invited to ascend when a door of light appeared. Those who choose to step through the door of light into eternity would be lifted from the third dimension into the fifth dimension. Ash discovered that this process of ascension described by Klein fitted into his previous scientific predictions. He used his vortex physics to spread the ascension message first in Australia and later around the world.[16]

Previously in 1986 Eric had channeled the ascended masters, and starting in 1988, he held weekly public meetings in his home. In January 1990 he was asked by Jesus to channel about the subject of ascension. It started in February and lasted six weeks. The messages were recorded

---

16. Ash, *Awaken*, chap.1, "Ascension," 17–23; Ash, *Superenergy and the Quantum Vortex*, 139.

on five cassette tapes, and they are known as the Ascension Tapes. According to Ash, the tapes reached thousands of lightworkers around the world.

In the tapes Jesus described the ascension process as follows:

So what is the process? You have been grown accustomed to the reincarnational experience—that is, having your soul incarnate in a body and going through a lifetime of a certain number of years, experiencing death and leaving your body to return again to the Earth. You have gone to some fourth dimensional areas. There are heavens and hells galore, with many experiences. Yet you always return to the body, for the body is the platform from which your launch will take place into the fifth dimension. This is the intention, let me say, of having a physical body, especially at this time. I would say that your ascension will be all but identical to my own. You will not leave your bodies behind and go to a higher state of consciousness. Your bodies will be transformed also. The molecules and atoms, your subatomic particles, all that you are will be transformed and accelerated into the fifth dimension. So you do not have to die. Well, that's some good news. Despite the fact that people have grown so accustomed to dying that it has become a common awareness or belief, I am telling you now, you do not have to die. And you will not.[17]

Klein's tapes and subsequent book predicted that prior to Earth changes, humans will have the opportunity to ascend into the light. Klein wrote: "A passage of light will appear, without any warning. Each of us will have the same opportunity to enter. Whether asleep or awake, we will all be able to pass through the door of light in full consciousness. The passage of light will be open for only a few moments. If we hesitate or choose not to enter, it will dissolve."[18]

---

17. Ash, *Awaken*, 18; see also Ash, *Superenergy and the Quantum Vortex*, 187–191 with a longer message; Klein, *The Crystal Chair*; Ash, *The Door of Light*, 27.
18. Klein, *The Crystal Chair*; Ash, *Awaken*, 18–19.

According to Ash, James Twyman also wrote about the Door to Eternity in *Emissary of Light* in 1996.[19] He wrote that by the light in our hearts, we will be drawn to the light of the Door to Eternity. We don't have to be afraid of anything and need worry about nothing. All we need to do is let go of our judgment of ourselves and others, let go of our fears and reservations, and trust the light. And surrender to it when it is offered to us.[20]

Ash wrote, "In the Eric Klein channelings it was made clear that the elect are not an elected few but are the few who elect to enter the door of light when it is offered to them. If the ascension predictions are true, entry into the fifth dimension would be a self-selection process. No heavenly agency is judging us. We, each of us, will judge ourselves."[21]

Compassion is the keyword of the Eric Klein channelings. Everyone gets the same chance out of compassion. In this way, people can avoid experiencing new traumas with the changes that will take place on Earth.[22] Next Ash describes the possible causes of the geophysical changes.[23]

In his book *Awaken*, David Ash wrote the following about ascension. "We will experience the collective ascension as if we were being lifted or beamed up into a safe, higher dimension."[24] Still, the decision to enter through the "portal of light"—or in Ash's words, "to step up in a vortex resonance teleportation beam"—takes courage. Ash said,

Crucial is the capacity of the person to surrender to a higher plan. This depends on the previous commitment to spiritual-human values over many lifetimes. Of course, it is difficult to prepare for

---

19. Ash, *Awaken*, 19n4
20. Ash, *Awaken*, 19.
21. Ash, *Awaken*, 19.
22. Ash, *Awaken*, 22.
23. Ash, *Superenergy and the Quantum Vortex*, 175–77 and 185–86 about the Earth's shifting crusts; Ash, *Awaken*, 47, 66–71 about Samvartaka of solar superstorms.
24. Ash, *Awaken*, 21.

ascension when no one knows for sure if or when it will happen. Ascension is about the total acceptance of everyone and everything. It is the manifestation of unconditional love. In the ascension process we lose nothing, but gain everything. Ascension and the changes of the Earth are not the end of the world, but the beginning of a New Golden Age of love and light. Ascension occurs in the eternal now when we practice unconditional love. Then will come the moment of ascension as the moisture evaporating off blades of grass when the sun rises.[25]

Ascension in the Twenty-First Century

Non-Western cultures previously predicted that after a period of great darkness a time would dawn when humanity would slowly awaken, starting in about 1875. For instance, traditional Hinduism has four different *yugas* or ages: the golden, silver, copper, and iron ages. This last iron age—also called the *Kali Yuga*—would have started around 3000 BCE and lasted until 1886.[26] The dates may vary slightly; for instance, some calculated it could be a period between 3001 BCE and 1899 CE.[27] In the Kali Yuga, humanity would have fallen asleep and forgotten who the human being in essence is. The Indian text entitled the *Vishnu Purana* predicted a spiritual awakening at the end of the downward spiral and a return to the golden age, but higher up the spiraling ladder.[28]

Meanwhile, the Native American peoples, including the Hopi and the Maya, saw a spiraling development with cycles related to the evolution of consciousness; their calendars predicted decline, chaos, and rise. In the Mayan calendar, the period of materialism, war, decay of norms, and decay of values and social structures lasted until 2012

25. Ash, *Awaken*, 23.
26. Ash, *Awaken*, 43, 58-60.
27. Harrie Salman, *Europe, A Continent with a Global Mission*, Sophia, Bulgaria: Kibea, 2009, 25.
28. Wyatt, *Cycles of Eternity*, 80–81.

because they counted in cycles of 5,125 years. The Mayans predicted in their long count that a fifth cycle began on August 11, 3114 BCE and ended on December 21, 2012. The new, sixth cycle began on December 22, 2012.

The Romanian author Radu Cinamar released a special book in 2019, in which he said, "What [Cezar Braz] told me agreed very well with what is currently known about raising the vibrational frequencies of the planet. The Earth is preparing to activate her higher dimensions, and that means primarily ascending into the subtle field that characterizes the etheric dimension, 4D. However, the 'old Earth' in the third dimension remains, with its limitations and vicissitudes. This divergence of vibrational frequency also inevitably leads to the growing apart of human beings; some will eagerly pursue spiritual evolution, while others—most—prefer to stick with what they already know, life on Earth in the 3D physical plane. This is a process that is already underway and will become more and more pronounced by 2020–2025, when certain galactic energy impulses occur with profound effects on human consciousness."[29]

Let's move on to look at these galactic energy impulses.

The Consciousness Revolution

Sharply increasing vibrational frequencies are emerging in a pattern that can be measured. For years I have been interested in the measurements of radiesthesia in places of power in the landscape. I wrote earlier about the old geomancy that works with the dowsing rod and new geomancy that works with modern techniques. A few years ago I came into contact with spiritual radiesthesia or dowsing with a pendulum.[30] In the following, I quote from articles on a website about spiritual radiesthesia.

---

29. Cinamar, *Forgotten Genesis*, chap. 2, sec. "The Motion of the Sphere Over Time in the Fourth Dimension (4-D) Creates a Kind of Tube, Like a 'Cane,'" n.p. Cesar Braz is a gifted and highly sensitive psychic from Romania.

30. Via DJR-Advies I came into contact with the measurements of Dineke Jongepier-Rijskamp; see the website of Dineke and Karel Jongepier.

Spiritual radiesthesia is based on universal but also scientific knowledge that all matter ultimately consists of frequency. Everything is vibration. Until recently, the vibrations outside the perceptible and knowable spectrum for the senses could not be measured. There were no measuring instruments for these increasingly refined vibrations. However, in the last century, the Frenchman André Bovis (1871–1947) developed a method to make a scale by means of radiesthesia to measure frequencies or wavelengths of both material and non-material things. These frequencies or wavelengths are shown on tables and can then be measured. In this way, transformation processes can be mapped. As a result, insights can be obtained that lie outside the spectrum perceptible to the senses. This opens up a world of dimensions, into which one is admitted only if the intermediary who makes use of the pendulum acts exclusively with pure motives.[31]

On December 21, 2012, the fifth long cycle of the Mayan calendar of 5,125 years ended. Not much seemed to change, but it did happen on a subtle level. Around that date, a shift in Bovis values took place in a certain segment of humanity. Those people who lived in connection with the god spark in the heart rose in vibration faster than those who lived materialistically and superficially and who felt little connection to their divine source, despite the outpouring of cosmic energies at the end of 2012. Before the end of 2012, both groups had a personal life vibration, or Bovis value of up to 8,500 units. But in those people who lived from the heart, after the impulse of December 21, 2012, values up to 10,500 were measured.[32] A second impulse followed on July 16, 2015, when values in people of goodwill rose to 23,000 units on the Bovis scale. Since then there have been several outpourings, with the largest occurring on March 17, 2021.

---

31. See the website of Dineke and Karel Jongepier. A book can also be measured on the Bovis scale, on which various dimensions are indicated. Then it can be measured whether a quote in a book or the book itself relates to a lower or higher world or even to the divine world. See measurements by DJR-Advies in chap. 2.
32. See DJR-Advies.

I learned the following on the aforementioned website. These outpourings of energy are actually outpourings of God's Holy Spirit. This means that the physical and etheric cell structure of billions of cells is changed from a carbon to a crystalline light structure by the ever-higher vibrations of life. The cells are more and more irradiated with cosmic light. It is a direct development process toward faith in God and God consciousness with which one realizes the conversion to a crystalline structure. The outpourings between December 21, 2012 and March 17, 2021 were portals to accelerated frequencies; it involved growth processes from a few thousand Bovis units to extraordinarily large numbers of units. You could label March 17, 2021, as a second birth or rebirth in a new phase of life.

Another impulse was measured on November 4, 2022. This time, the central sun sent out a major impulse to Earth. I received this message by telephone and read about it later on the aforementioned website. It read:

In the night of November 3 to 4, 2022, there was a large outflow of energy from the central sun in our universe. This outflow has caused that, if one is well attuned to the cosmic principle of love, from that moment on, among other things, an activation was initiated to restore our 12-strand DNA, of which only 2 strands have been operational for thousands of years. The rest are called junk DNA by science. However, when our 12 strands are restored, we become a whole person with unprecedented possibilities of expression. It is very remarkable that the Swiss clairvoyant and author Christina von Dreien also refered to the impulse of the central sun in her November newsletter, without knowing the exact date of the outflow of the central sun.[33]

On Tuesday, November 15, 2022, I received a digital newsletter from Christina von Dreien, which, after translation from German, contained the following message:

---

33. See DJR-Advies website, Messages uploaded on November 18, 2022, on the homepage after measurement on November 4, 2022.

The vibrational raising and ascension is not only taking place on Earth, but also in our solar system. The other planets and the sun are also in such a process. The sun we see in our sky is energetically connected to the central sun. Through the sun we can receive energetic information that will help us open our consciousness more and reactivate our DNA. When our DNA is fully active again, we will again be able to use skills such as telepathy and telekinesis. These skills belong to people. Normally people should have these abilities. Whether a person wakes up in this incarnation and reactivates the DNA depends on his/her consciousness. Some people cannot yet wake up in this incarnation. But whoever can, will. With their awareness they will enable the great change for the good and thereby also help that the changes that are needed for it will take place more gently.

It is remarkable that the spiritual radiesthesia measurement confirmed Christina von Dreien's observation without there having been mutual contact. Portals open for someone with an open and warm heart.

Jesus said in logion 61 of the Gospel of Thomas:

> *For this reason I say,*
> *if one is whole,*
> *one will be filled with light,*
> *but if one is divided,*
> *one will be filled with darkness.*

21

A COLLECTIVE ASCENSION

After learning about ascension on an individual level, it is time to examine ascension on a collective level. It is Alariel who informs us about a collective ascension and I paraphrase and summarize his words: The whole galaxy is rising in vibration, ascending into the Light and returning to the Source. This is a major shift in emphasis within all creation as God stops moving outwards and down into matter and starts moving inwards and up into the Light. The whole pattern within the universe is beginning to reflect this shift, and the changes all are experiencing on the Earth need to be seen as part of this bigger picture. The time has come for individual ascension, planetary ascension and galactic ascension into a higher and subtler state of being. This is the context within which all the processes of change at the personal and planetary level occur.[1]

To frequence your consciousness is to uplift its frequency in a series of wave-like modulations. Initially, your consciousness can only occasionally visit and touch that level, but with each impulse the ability to hold the higher frequencies increases. Your consciousness moves upward in a series of pulses that lift its vibrational frequencies. These pulses are like waves; they raise the frequency to higher and higher levels, with a drop-off at times, but then in a continuous upward movement. As a result, your consciousness is renewed, you develop new skills and also a new vocabulary to express new experiences. In these vibrations from

---

1. Wilson and Prentis, *Atlantis and the New Consciousness*, 111–12.

higher planes, not only newborn children go along; it is also the elderly who have time after retirement to develop their awareness and sensitivity. They also contribute to the development of this new language. Together with the children of the New Age, they will develop human consciousness into new realms that can only be called a revolution in consciousness.[2]

With expanding consciousness, the DNA transforms. It rises in vibration; compare it to going up a scale. Three elements form a threefold whole and merge together, namely: consciousness, DNA and vibration. It is about a holistic system that moves through a vortex of change. In addition, the individual increase in consciousness contributes to an increase in the collective human consciousness. It is about developing an inner consciousness from within, not a machine-driven process from the outside. As human consciousness expands, a technology of light is revealed. The effect of the light within the consciousness and the unconditional love in the heart leads to a full activation of the light body.[3]

The large amounts of light generate dosed keys in your DNA. This sends you a wake-up call and intensifies your energy field. That affects your etheric body and stimulates every cell within your system to vibrate faster. Your nervous and glandular system must adapt to this and be restructured. Compared to your grandparents, you have already progressed a lot. The real big change comes with the transition from limited third dimensional or 3D consciousness to full fifth dimensional or 5D consciousness.[4]

The Light Body is essentially the vehicle of ascension. However, many lightworkers have a Light Body that is not yet fully activated. It is activated most effectively through focusing the energy of Unconditional Love through the heart. This needs to be Love both for others and for oneself, and this is a balance that some may find difficult. When you move through this activation process, you will be experiencing exactly

---

2. Wilson and Prentis, *Atlantis and the New Consciousness*, 138.
3. Wilson and Prentis, *Atlantis and the New Consciousness*, 140.
4. Wilson and Prentis, *Beyond Limitations*, 78.

the process that Yeshua (and Miryam) were trying to teach. The Light Body activation causes pulses of Light which expand the consciousness and integrate the system of force-centers in the body into a new continuum of expanded Being. When the Light Body is activated, it sends a series of Light pulses into the personality to clarify and integrate the whole system of chakras or force-centers, reattuning this system so it becomes a platform for expanding consciousness and rapid spiritual growth.[5] Now the chakras are starting to unify, you become a crucible within which spiritual transformation can occur.[6] When it is fully absorbed and integrated into the whole being right down to the cellular level, this activation leads to the beginning of an ascension process that will open access to further levels of spiritual growth.[7]

Ascension shifts the focus permanently to the next dimensional reality. This does not necessarily mean leaving the Earth, but it does mean entering a new level of empowerment and a fresh commitment to a life of loving service. Because the Earth has already risen so much in vibration, the option of staying here to serve in an ascended state has now become available. This is a very different situation to that of two thousand years ago when full ascension inevitably meant disappearing into the Light and leaving the Earth's frequency, since that frequency was so dense and so far from the vibration of the ascended state of being.[8]

The whole transition upwards into the Light needs to be put in a greater context of Earth's place in the universe. The presence of the New Age children demonstrates that larger context. Their presence here, which is essentially a gift from other star civilizations and galaxies underlines the Brother and Sisterhood of all Sentient Beings. You are not alone in moving upwards into the Light and at this vital time you need and are receiving vast amounts of help.[9]

---

5. Wilson and Prentis, *Atlantis and the New Consciousness*, 140–41.
6. Wilson and Prentis, *Atlantis and the New Consciousness*, 118.
7. Wilson and Prentis, *Atlantis and the New Consciousness*, 141.
8. Wilson and Prentis, *Atlantis and the New Consciousness*, 141.
9. Wilson and Prentis, *Atlantis and the New Consciousness*, 141–42.

When you are in contact with personal guides, they will be very active at this time, and the whole angelic realm stands ready to help in any way you find appropriate. All you have to do is ask! There are Star Masters from a whole range of civilizations within your galaxy: especially Sirius, the Pleiades, Arcturus, Orion, and Vega.[10] And Star Masters from the galaxy of Andromeda will become increasingly important as you focus more and more on Oneness: they are the supreme teachers of Unity Consciousness and how it can be practically applied. Help is also available through the Ashtar Command: this brings together Star Beings from many civilizations focusing a wide range of multidimensional knowledge and experience.[11] In your separation you forgot that everything in the universe is connected.

All physical sentient beings share in the Divine Image, the mathematical and geometric basis on which your forms are created. The mathematics and geometry provide the template upon which the angel host creates the architecture of form for sentient beings throughout the universe. The underlying foundation of mathematics and geometry connects the whole structure into a single holistic unit. But there is a still deeper unifying element, and that is the foundational Energy of Unconditional Love. It was Love that brought the universe into being, and Love that sustains it and provides the impetus for change within the consciousness of the sentient beings. Whatever spiritual path you choose, the transformative energy of Unconditional Love is there to support and nourish that path and carry through the process of change and transformation within the heart. Simply attuning to this Energy and invoking it within the heart will begin a process of profound change.[12]

When you consider the power of Love working together with the vast resources of the Technology of Light, you begin to glimpse the whole arc of the scheme of spiritual evolution. What we're really talking about here is the power of the Spirit. If you define magic as a process

---

10. Wilson and Prentis, *Beyond Limitations*, 79.
11. Wilson and Prentis, *Beyond Limitations*, 79.
12. Wilson and Prentis, *Atlantis and the New Consciousness*, 142.

that transcends the law of physics then as the Spirit is certainly capable of transcending those laws, its functioning can be seen as essentially magic. When the Spirit moves, it moves in such a multidimensional way that 3D space and linear time are transcended and anything is possible. The Spirit links together all beings and transcends all the dimensions of space and frequencies of time. Spirit affirms Oneness of Life, Oneness of Being, Oneness of Essence.[13]

When Yeshua said that Mary Magdalene was "the Woman who understands the All," this was what he meant—an understanding and knowing and living at the level where there is simply One Energy and One Consciousness uniting all that is. And that Energy, that Consciousness, is Unconditional Love.[14]

Native American peoples speak of living lightly on the Earth, light in the sense of living lightly and moving with the flow of changes, without clinging too much to the old.[15] Move through the physical-emotional-mental-spiritual continuum. It is essential to bring your ideals and aspirations down into practical manifestations at the physical level.

Sadly, the Essene experiment (of Yeshua and Miryam) could not be sustained and for a time the Light passed out of human memory and human experience. But now, two thousand years later, another great opportunity to manifest the Light emerges. As your planet moves upward in vibration, many human beings are experiencing a process of transformation and awakening. Once again there is closer contact with the angelic realm. This is High Magic, the magic of the Spirit, through which all hurts are healed and all things are made new. This is the opening of the door to a multidimensional experience as humanity awakens and your planet becomes a sacred planet within a galaxy that is returning to Source. This is the beginning of a new life in which the Children of Light return from their wanderings and rise into the wonder of everlasting joy.[16]

---

13. Wilson and Prentis, *Atlantis and the New Consciousness*, 142–43.
14. Wilson and Prentis, *Atlantis and the New Consciousness*, 145.
15. Wilson and Prentis, *Beyond Limitations*, 78.
16. Wilson and Prentis, *Atlantis and the New Consciousness*, 147.

THE CIRCLE IS COMPLETE

Back to the initial question at the beginning of this book. In chapter 2, I asked myself, "Do ancient authors' accounts match eye-witness accounts of people in regression? Do the latter add anything? Do we get a clearer, more complete, and more human picture of that time? If so, then it is desirable not to exclude this material anymore." I promised: "In the Conclusion at the end of the book I will give the end result of this challenging exercise." You, dear reader, will have to feel and experience and decide for yourself whether you consider the information provided by the alternative sources reliable and whether it resonates with you; the alternative sources cannot be rationally verified or proven, but require trust.

Nevertheless, numerous historical sources point in the direction indicated by the alternative sources. In my opinion, the historical sources largely confirm the alternative sources. Indeed, journeys to France, England, and Central Asia appear to have been made by Joseph of Arimathea, Mary Magdalene, and Yeshua. All this is not as out of the blue as it may seem at first glance. It seems that the information provided by the alternative sources is reliable, since they are supported by data from historical sources. In addition, the alternative sources provide added value. In my opinion, they make our worldview more insightful, more complete, more humane, and more universal. We, the Children of Light, are ready for new knowledge—knowledge for the New Millennium.

Completing the research and writing this book has given me much satisfaction and made me deeply grateful. You realize that you are a small link in a large renewal movement in which transformation is the key word.

In logion 50, of the Gospel of Thomas Jesus said,

> *If they say to you,*
> *"Where did you come from?"*
> *Say to them:*
> *"We have come from the light,*
> *from the place where the light came into being by itself."*

BIBLIOGRAPHY

Ahmad, Khwaja Nazir. *Jesus in Heaven on Earth: Journey of Jesus to Kashmir: His Preaching to the Lost Tribes of Israel and Death and Burial in Srinagar.* Columbus, OH: Ahmadiyya Anjuman Isha' at Islam Lahore Inc., 1998. First published in 1952 by Muslim Mission & Literary Trust, Woking, England, 1952.

Anonymous. *The Crucifixion—By an Eye-Witness: A Letter Written Seven Years after the Crucifixion by a Personal Friend of Jesus in Jerusalem to an Esseer Brother in Alexandria.* Chicago: Indo-American Book Company, 1907.

Ash, David. *Activation for Ascension.* Cape Town, South Africa: Kima Global Publishing, 1995.

———. *Awaken: The Physics of Ascension,* 2nd ed. Cape Town, South Africa: Kima Global Publishers, 2020.

———. *Continuous Living in a Living Universe.* Cape Town, South Africa: Kima Global Publishing, 2015.

———. *The New Physics of Consciousness: Reconciling Science and Spirituality.* Cape Town, South Africa: Kima Global Publishing, 2007.

———. *The New Science of the Spirit.* London: The College of Psychic Studies, 1995.

———. *The Role of Evil in Human Evolution: Exposing the Dark to Light.* Cape Town, South Africa: Kima Global Publishing, 2007.

———. *Superenergy and the Quantum Vortex.* Cape Town, South Africa: Kima Global Publishing, 2023.

———. *The Vortex Theory: The Bridge between Ancient Yoga and Modern Physics.* Cape Town, South Africa: Kima Global Publishing. 2015.

Ash, David, and Anna Ash. *The Tower of Truth.* Dolsdon: self-published, D. and A. Ash, 1977.

Ash, David, and Peter Hewitt. *The Vortex: Key to Future Science.* Bath, Somerset: England: Gateway Books, 1990.

Baigent, Michael, Richard Leigh, and Henry Lincoln. *Holy Blood, Holy Grail.* New York: Dell, 1983.

Baring, Anne. *The Dream of the Cosmos: A Quest for the Soul.* Dorset, England: Archive Publishing, 2013.

Baring-Gould, Sabine. *A Book of Cornwall.* London: Methuen & Company Limited, 1923.

Barker, Margaret. "Beyond the Veil of the Temple: The High Priestly Origins of the Apocalypses." *Scottish Journal of Theology* 51, no. 1 (February 1998): 1–21.

———. *Christmas: The Original Story.* London: Society for Promoting Christian Knowledge, 2008.

———. *An Extraordinary Gathering of Angels.* London: MQ Publications Ltd., 2004.

———. *The Gate of Heaven: The History and Symbolism of the Temple in Jerusalem.* Sheffield, England: Sheffield Phoenix Press, 2008.

———. *The Great Angel: A Study of Israel's Second God.* London: Society for Promoting Christian Knowledge, 1992.

———. *The Great High Priest: The Temple Roots of Christian Liturgy.* London: T&T Clark, 2003.

———. *The Great Lady: Restoring Her Story.* Sheffield, England: Sheffield Phoenix Press, 2023.

———. "The Images of Mary in the Litany of Loreto." *Usus Antiquior* 1, no. 2 (July 2010): 110–131.

———. *King of the Jews: Temple Theology in John's Gospel.* London: Society for Promoting Christian Knowledge, 2014.

———. *The Lost Prophet: The Book of Enoch and Its Influence on Christianity.* Sheffield, England: Phoenix Press, 2005.

———. *The Mother of the Lord, Vol 1: The Lady in the Temple.* London: Bloomsbury, 2012.

———. *The Older Testament: A Survival of Themes From the Ancient Royal Cult in Sectarian Judaism and Early Christianity.* Sheffield, England: Phoenix Press, 2005.

———. *Revelation of Jesus Christ, Which God Gave to Him to Show to His Servants What Must Soon Take Place (Revelation 1.1.)* Edinburgh: T&T Clark, 2000.

———. *Temple Mysticism: An Introduction.* London: Society for Promoting Christian Knowledge, 2011.

———. *Temple Themes in Christian Worship.* London: T&T Clark, 2007.

———. "Text and Context." Papers, Margaret Barker's website. 2002.

Beavis, Mary Ann, and Ally Kateusz, eds. *Rediscovering the Marys: Maria, Mariamne, Miriam.* London: T&T Clark, 2020.

Begg, Ean. *The Cult of the Black Virgin.* New York: Arkana, 1996.

Bohm, David. *Wholeness and the Implicate Order.* New York: Routledge Classics, 1980.

Broothen, Bernadette. *Women Leaders in the Ancient Synagogue: Inscriptional Evidence and Background Issues.* Chico, CA: Brown Judaic Studies, 2020.

Brown, Raymond E. *The Community of the Beloved Disciple: The Life, Loves and Hates of an Individual Church in New Testament Times.* New York: Doubleday & Company, 1979.

———. *The Death of the Messiah, From Gethsemane to the Grave: A Commentary on the Passion Narratives in the Four Gospels.* 2 vols. New York: Doubleday & Co., 1993.

———. *The Gospel According to John (I–XII).* New York: Doubleday & Co., 1966.

———. *The Gospel According to John (XIII–XXI).* New York: Doubleday & Co., 1970.

———. *The Gospel and Epistles of John: A Concise Commentary.* Collegeville, MN: Liturgical Press, 1988.

———. "Roles of Women in the Fourth Gospel." *Theological Studies* 36, no. 4 (December 1975): 688–699.

Brown, Raymond E., and Raymond F. Collins, eds. *Canonicity: The New Jerome Biblical Commentary.* Englewood Cliffs, NJ: Prentice Hall, 1990.

Cannon, Dolores. *Jesus and the Essenes.* Huntsville, AR: Ozark Mountain Publishing, 2009.

Capt, E. Raymond. *The Traditions of Glastonbury.* Thousand Oaks, CA: Artisan Sales Edition, 1983.

Cerio, Joan. *Hardwired to Heaven: Download Your Divinity through Your Heart and Create Your Deepest Desires.* Rochester, VT: Findhorn Press, 2014.

Charles, R. H. *The Book of Enoch.* With an Introduction by W. O. E. Oesterley. Mineola, NY: Dover, 2007.

Charlesworth, James H. *The Beloved Disciple: Whose Witness Validates the Gospel of John?* Valley Forge, PA: Trinity Press International, 1995.

Charlesworth, James H., ed. *The Old Testament Pseudoepigrapha.* 2 vols. Peabody, MA: Hendrickson, 2011 (2nd ed.)

Cinamar, Radu. *Forgotten Genesis*. New York: Sky Books, 2020.

———. *Transylvanian Moonrise: A Secret Initiation in the Mysterious Land of the Gods*. New York: Sky Books, 2021.

Common Bible: Revised Standard Version with the Apocrypha/Deuterocanonical Books, The. Reprint. New York, Glasgow, Toronto, Sydney, and Auckland: Collins, 1995.

Creme, Benjamin. *The Ageless Wisdom Teaching*. London: Share International Foundation, 1996.

Dale, Cyndi. *The Subtle Body: An Encyclopedia of Your Energetic Anatomy*. Boulder, CO: Sounds True, 2009.

de Boer, Esther A. *The Gospel of Mary: Beyond a Gnostic and a Biblical Mary Magdalene*. New York: Bloomsbury, 2004.

———. "Mary Magdalene and the Disciple Jesus Loved," *European Electronic Journal for Feminist Exegesis*, 1 (2000). Lectio Difficilior website.

de Quillan, Jehanne. *The Gospel of the Beloved Companion: The Complete Gospel of Mary Magdalene*. Translation and Commentary. Foix: Athara Éditions, 2010.

Dietzfelbinger, Konrad. *Mysteriescholen*. Deventer, Netherlands: Ankh Hermes, 2000.

Dobson, Cyril C. *Did Our Lord Visit Britain as They Say in Cornwall and Somerset?* London: Covenant Publishing Co., 1999.

Dunford, Barry. *The Holy Land of Scotland: Jesus in Scotland and the Gospel of the Grail*. Perthshire, Scotland: Sacred Connections, 2002.

———. *Vision of Albion: The Key to the Holy Grail; Jesus, Mary Magdalene and the Christ Family in the Holy Land of Britain*. Perthshire, Scotland: Sacred Connections, 2008.

Faber-Kaiser, Andreas. *Jesus Died in Kashmir*. London: Gorden & Cremonesi, 1977.

Faillon, Etienne Michel, ed. *Monuments Inédits Sur L'apostolat De Sainte Marie-Madeleine En Provence Et Sur Les Autres Apôtres De Cette Contrée, Saint Lazare, Saint Maximin, Sainte Marthe Et Les Saintes Maries Jacobé Et Salomé*, Vol. II. Sydney: Wentworth Press, an imprint of Creative Media Partners, 2011. First published 1848 by Aux Ateliers Catholiques de Petit-Montrouge, Paris.

Freedman, David Noel, Allen C. Myers, and Astrid B. Beck, eds. *Eerdmans Dictionary of the Bible*. Grand Rapids, MI: W. B. Eerdmans Publishing Co., 2000.

Friedman, Norman. *Bridging Science and Spirit: Common Elements in David*

Bohm's Physics, the Perennial Philosophy and Seth. Eugene, OR: The Woodbridge Group, 1997.

Gardner, Laurence. *Bloodline of the Holy Grail: The Hidden Lineage of Jesus Revealed.* Shaftsbury, England: Element Books, 1996.

———. *The Grail Enigma: The Hidden Heirs of Jesus.* London: HarperElement, 2008.

———. *The Magdalene Legacy: The Jesus and Mary Bloodline Conspiracy.* London: HarperElement, 2005.

Gaster, Theodor H., ed. *The Dead Sea Scriptures.* In English Translation with Introduction and Notes by Theodor H. Gaster. 3rd ed. New York; Anchor Books, 1976.

Goldberg, Shari. "The Two Choruses Become One: The Absence/Presence of Women in Philo's 'On the Contemplative Life.'" *Journal for the Study of Judaism*, 39 (2008): 459–70.

Goranson, Stephen. "On the Hypothesis that Essenes Lived on Mt. Carmel." *Revue de Qumran* 4, no. 36 (December 1978): 563–567.

Goswami, Amit. *The Self-Aware Universe: How Consciousness Creates the Material World.* New York: Tarcher Pedigree, 1993.

Gottheil, Richard, and H. G. Enelow. "Arnold of Cîteaux." Jewish Encyclopedia website.

Greene, Elizabeth. "The Celestial Ascent of the Soul: The Morphology of an Enduring Idea." Master's thesis, Bath Spa University, 2006. CosmoCritic website, The 9th House.

Grossinger, Richard. "An Unbottomable Void in Mind." *The Paradigm Explorer, Journal of the Scientific and Medical Network*, 137 (2021): 16–18.

———. *Bottoming Out the Universe: Why There Is Something Rather Than Nothing.* Rochester, VT: Park Street Press, 2020.

Hanegraaff, Wouter, ed. *Dictionary of Gnosis and Western Esotericism.* Leiden, Netherlands: Brill, 2006.

———. *New Age Religion and Western Culture: Esotericism in the Mirror of Secular Thought.* Leiden, Netherlands: Brill, 1996.

Hartley, Christine. *The Western Mystery Tradition. The Esoteric Heritage of the West,* Wellingborough, UK: Aquarian, 2004.

Hassnain, Fida. *A Search for the Historical Jesus: From Apocryphal, Buddhist, Islamic and Sanskrit Sources.* Bath, UK: Gateway Books, 2004.

Hassnain, Fida, and Levi Dahan. *The Fifth Gospel: New Evidence from the Tibetan, Sanskrit, Arabic, Persian and Urdu Sources About the Historical Life of Jesus*

Christ After the Crucifixion. London: Blue Dolphin Publishing, 2007.

Hawkins, David R. *Power Versus Force: The Hidden Determinants of Human Behavior*. Carlsbad, CA: Hay House, 2002.

Heartsong, Claire. *Anna, Grandmother of Jesus: A Message of Wisdom and Love*. Carlsbad, CA: Hay House, 2017.

Heartsong, Claire, and Catherine Ann Clemett. *Anna, the Voice of the Magdalenes: A Sequel to Anna, Grandmother of Jesus*. Carlsbad, CA: Hay House, 2017.

Hunt, Valerie. *Mind Mastery Meditations: A Workbook for the "Infinite Mind."* Malibu, CA: Malibu Publishing Co. 1997.

Hennecke, Edgar. *Neutestamentliche Apocryphen in deutscher Übersetzung, bd 1: Evangelien, bd 2: Apostolisches, Apocalypsen und Verwandtes*. Edited by Wilhelm Schneemelcher. Tübingen, Germany: Mohr Seibeck, 1964.

Jones, Kathy. *In the Nature of Avalon: Goddess Pilgrimages in Glastonbury's Sacred Landscape*. Glastonbury, UK: Ariadne Publications, 2007.

———. *Priestess of Avalon, Priestess of the Goddess: A Renewed Spiritual Path for the 21st Century*. Glastonbury, UK: Ariadne Publications, 2006.

Josephus, Flavius. *Antiquities of the Jews*. Cambridge, MA: Harvard University Press, 1966.

———. *The Jewish War*. London: William Heinemann, 1926.

Jowett, George J. *The Drama of the Lost Disciples*. 12th ed. London: Covenant Publishing Co. Ltd., 1993.

Jusino, Ramon. "Mary Magdalene: Author of the Fourth Gospel?" Master's thesis. beloveddisciple website, 1998.

Kastrup, Bernardo. *Science Ideated: The Fall of Matter and the Contours of the Next Mainstream Scientific Worldview*. Winchester, UK: Iff Books, 2020.

Kateusz, Ally. *Mary and Early Christian Women: Hidden Leadership*. London: Palgrave Macmillan, 2019.

Klijn, Albertus Frederik Johannes. *Apocriefen van het Nieuwe Testament I, vert., ingeleid en toegelicht onder eindred. v. A.F.J. Klijn*. Kampen, Netherlands: Kok, 1984.

———. *Apocriefen van het Nieuwe Testament II, vert., ingeleid en toegelicht onder eindred. v. A.F.J. Klijn*. Kampen, Netherlands: Kok, 1985.

Kersten, Holger. *Jesus Lived in India*. Shaftsbury, England: Element Books, 1994.

Kersten, Holger, and Elmar R. Gruber. *The Jesus Conspiracy: The Turin Shroud and The Truth About the Resurrection*. Shaftsbury, England: Element Books, 1994.

Klein, Eric. *The Crystal Chair: A Guide to the Ascension*. Blue Hill, ME: Medicine Bear Hill, 1992.

Klink, Joanne. *De Onbekende Jezus*. Baarn, Netherlands: Ten Have, 1996.

———. *Vroeger toen ik groot was. Vergaande herinneringen van kleine kinderen*. Baarn, Netherlands: Ten Have, 1990.

Kovács, Betty. *Merchants of Light: The Consciousness that Is Changing the World*. Claremont, CA: The Kamlak Center, 2019.

Larson, Martin A. *The Essene-Christian Faith: A Study in the Sources of Western Religion*. New York: Philosophical Library, 1980.

———. *The Religion of the Occident*. New York: Philosophical Library, 1959.

Lewis, Henry Ardern. *Christ in Cornwall and Glastonbury the Holy Land of Britain*. Glastonbury, England: W. H. Smith & Son, 1946.

Lewis, Lionel Smithett. *St. Joseph of Arimathea at Glastonbury or the Apostolic Church of Britain*. Glastonbury, England: Avalon Press, 1922. Reprinted. Cambridge, England: Lutterworth Press, 2004.

Lorimer, David. Review of *Bottoming Out the Universe*, by Richard Grossinger *Paradigm Explorer*, 137, part 3 (2021): 53–54.

———. "The Gospel of the Beloved Companion." Science and Nonduality website, August 22, 2019.

Luttikhuizen, Gerard P. *De evangeliën van Thomas, Maria Magdalena en Judas: vroeg-christelijke esoterie en gnosis*. Almere, Netherlands: Parthenon, 2018.

———. *De veelvormigheid van het vroegste christendom*. 3rd ed. Budel, Netherlands: Damon VOF, 2005.

———. *Gnostische Geschriften I, Het Evangelie naar Filippus, de brief van Petrus aan Filippus*. Kampen, Netherlands: Kok, 1988.

———. "Monism and Dualism in Jewish-Mystical and Gnostic Ascent Texts." In *Dead Sea Scrolls and Other Early Jewish Studies in Honour of Florentino García Martínez*, edited by Anthony Hilhorst, Anthony, Émile Puech, and E. Tigchelaar. Vol. 122 of *Supplements to the Journal for the Study of Judaism*.

Mann, Nicholas R. *The Isle of Avalon: Sacred Mysteries of Arthur and Glastonbury*, rev. ed. London: Green Magic, 2001.

McGilchrist, Iain. *The Master and his Emissary: The Divided Brain and the Making of the Western World*. New Haven, CT: Yale University Press, 2009.

———. *The Matter with Things: Our Brains, Our Delusions and the Unmaking of the World*. 2 vols. London: Perspectiva Press, 2021.

McTaggart, Lynn. *The Field: The Quest for the Secret Force of the Universe.* New York: Quill, 2001.

Mehler, Stephen S. *From Light into Darkness: The Evolution of Religion in Ancient Egypt.* Kempten, IL: Adventures Unlimited Press, 2005.

———. *The Land of Osiris: An Introduction into Khemitology.* Kempten, IL: Adventures Limited Press, 2001.

Meyer, Marvin W., ed. *The Ancient Mysteries: A Sourcebook of Sacred Texts.* Philadelphia: University of Pennsylvania Press, 1999.

———. *The Gospels of Mary: The Secret Tradition of Mary Magdalene, the Companion of Jesus.* San Francisco: HarperOne, 2004.

———. *Nag Hammadi Scriptures. The Revised and Updated Translation of Sacred Gnostic Texts Complete in One Volume.* International edition. New York: HarperCollins, 2007.

Michell, John. *New Light on the Ancient Mystery of Glastonbury.* Glastonbury, UK: Gothic Image Publications, 1997.

———. *The New View Over Atlantis.* London: Thames & Hudson, 1983.

Mycoff, David. *The Life of Saint Mary Magdalene and of Her Sister Saint Martha: A Medieval Biography.* Translated and annotated by David Mycoff. Kalamazoo, MI: Cistercian Publications, 1989.

Nestle-Aland. *Novum Testamentum Graece et Latine.* Stuttgart: Deutsche Bibelgesellchaft, 1984.

Notovitch, Nicholas. *The Unknown Life of Jesus Christ.* Radford, VA: Wilder Publications, 2008.

Noyce, John. *The Inner Ascent.* Vol. 21 of *History Enlightened.* Self-published, Lulu, 2018.

Olsson, Suzanne. *Jesus in Kashmir, the Lost Tomb.* Self-published, 2007.

Philo. *De Vita Contemplativa.* Translated by F. H. Colson. Loeb Classical Library 363. Cambridge, MA: Harvard University Press, 1985.

Picknett, Lynn. *Mary Magdalene: Christianity's Hidden Goddess.* London: Constable & Robinson, 2000.

Pliny the Elder. *Historia Naturalis, Vol. I: Books 1–2 Natural History II. Translated by H. Rackham.* Loeb Classical Library 330. Cambridge, MA: Harvard University Press, 1942.

Prophet, Elizabeth Clare. *The Lost Years of Jesus: On the Discoveries of Notovitch, Abhedananda, Roerich and Caspari.* Missoula, MT: Summit University Press, 1984.

———. *Mary Magdalene and the Divine Feminine—Jesus' Lost Teachings on*

Woman: How Orthodoxy Suppressed Jesus' Revolution for Woman and Invented Original Sin. Missoula, MT: Summit Publications, 2005.

Quispel, Gilles. *Corpus Hermeticum*. Amsterdam: In de Pelikaan, 1990.

———. *Het evangelie van Thomas uit het Koptisch vertaald en toegelicht*. Amsterdam: In de Pelikaan, 2005.

Ralls, Karen. *Mary Magdalene: Her Mysteries and History Revealed*. New York: Shelter Harbor Press, 2013.

Ramacharaka, Yogi. *Advanced Course in Yogi Philosophy and Oriental Occultism*. Chicago: The Yogi Publication Society, 1905. Reprinted. New York: Cosimo Classics Philosophy, 2005.

Rameijer, Jaap. *Maria Magdalena in Frankrijk*. Soesterberg, Netherlands: Uitgeverij Aspekt, 2013.

Ransome, Hilda M. *The Sacred Bee in Ancient Times and Folklore*. London: George Allen & Unwin, 1937. Reprinted. North Chelmsford, MA: Courier Corporation, 2004.

Rahtz, Philip. *Glastonbury*. London: Batsford Ltd., 1993.

Rennison, Susan Joy. *Afstemmen van op de Kosmos: Elektromagnetisme en spirituele evolutie*. Deventer, Netherlands: Ankh Hermes, 2010.

Rigoglioso, Marguerite. *The Cult of Divine Birth in Ancient Greece*. New York: Palgrave Macmillan, 2009.

———. *The Mystery Tradition of Miraculous Conception: Mary and the Lineage of Virgin Births*. Rochester, VT: Inner Traditions, 2021.

———. *Virgin Mother Goddesses of Antiquity*. New York: Palgrave Macmillan, 2010.

Red Star, Nancy. *Star Ancestors: Extraterrestrial Contact in the Native American Tradition*. Rochester, VT: Bear and Company, 2012.

Roberts, Jane. *The Seth Material*. Toronto: Bantam Books, 1970.

Robinson, James M., ed. *The Nag Hammadi Library: The Definitive Translation of the Gnostic Scriptures; Complete in One Volume*. San Francisco: Harper, 1990.

Scholem, Gershom. *Origins of the Kabbalah*. Princeton: The Jewish Publication Society, 1987.

Schonfield, Hugh. *The Passover Plot: New Light on the History of Jesus*. London: Hutchinson, 1965. Reprinted. London: The Hugh & Helene Schonfield World Service Trust, 2017.

Slavenburg, Jacob, and Willem Glaudemans, eds. *De Nag Hammadi Geschriften: Een integrale vertaling van alle teksten uit de Nag Hammadi Codices en de Berlijnse Codex*. Deventer, Netherlands: Ankh Hermes, 2004.

Southern, R. W. *Western Society and the Church in the Middle Ages.* New York, Penguin Books: 1990.

Starbird, Margaret. *The Woman with the Alabaster Jar: Mary Magdalene and the Holy Grail.* Rochester, VT: Bear & Company, 1993.

Szekely, Edmond Bordeaux. *The Discovery of the Essene Gospel of Peace: The Essenes and the Vatican.* San Diego: Academy Books, 1977.

———. *The Essene Gospel of Peace: The Third Century Aramaic Manuscript and old Slavonic Texts.* Compared, edited, and translated by Edmond B. Szekely. n.p.: AudioEnlightenment, 2018

Taylor, John William. *The Coming of the Saints: Imaginations and Studies in Early Church History and Tradition.* London: Methuen and Co., 1906. Reprinted by Franklin Classics, an imprint of Creative Media Partners, 2018.

van den Broek, Roelof. *De Taal van de Gnosis: Gnostische teksten uit Nag Hammadi.* Baarn, Netherlands: Ambo, 1986.

———. *Gnosis in de Oudheid: Nag Hammadi in Context.* Amsterdam: In de Pelikaan, 2010.

———. *Gnostic Religion in Antiquity.* Cambridge, UK: Cambridge University Press, 2013.

van den Broek, Roelof, and Gilles Quispel, eds. *Hermetische Geschriften. Ingeleid, vertaald en toegelicht door R. van den Broek en G. Quispel.* Amsterdam: In de Pelikaan, 2016.

van den Broek, Roelof, and Cis van Heertum. *From Poimandres to Jacob Böhme: Gnosis, Hermetism and the Christian Tradition.* Amsterdam: In de Pelikaan, 2000.

van der Meer, Annine. *The Black Madonna: From Primal to Final Times.* The Hague, Netherlands: Pansophia Press, 2020.

———. *The Language of MA the Primal Mother: The Evolution of the Female Image in 40,000 Years of Global Venus Art.* The Hague, Netherlands: Pansophia Press, 2013.

———. *Mary Magdalene Unveiled: Hidden Sources Restore Her Broken Image.* The Hague, Netherlands: Pansophia Press, 2013.

van der Toorn, Karel, Bob Becking, and Pieter W. van der Horst, eds. *Dictionary of Deities and Demons in the Bible.* Leiden: Brill, 1995.

van Dijk, Danielle. *Maria Magdalena, de Lady van Glastonbury and Iona: Over de oudste wortels van het Keltische christendom.* Amsterdam: Cichorei, 2018.

van Oyen, Paul. *De Vergeten Jaren van Jezus in India en Tibet.* Amsterdam: Conversion Productions, 2010.

de Voragine, Jacobus. *La Légende Dorée*. Paris: Éditions du Seuil, 1998.

de Vries, Marja. *The Whole Elephant Revealed: Insights into the Existence and Operation of Universal Laws and the Golden Ratio*. Essex, UK: Axis Mundi Books, 2012.

Wakefield, Walter L., and Austin P. Evans. *Heresies of the High Middle Ages*. New York: Columbia University Press, 1969.

Wilson, Stuart, and Joanna Prentis. *Atlantis and the New Consciousness*. Huntsville, AR: Ozark Mountain Publishing, 2012.

———. *Beyond Limitations: The Power of Conscious Co-Creation*. Huntsville, AR: Ozark Mountain Publishing, 2012.

———. *The Essenes: Children of the Light*. Huntsville, AR: Ozark Mountain Publishing, 2021.

———. *The Magdalene Version: Secret Wisdom from a Gnostic Mystery School*. Huntsville, AR: Ozark Mountain Publishing, 2019.

———. *Power of the Magdalene: The Hidden Story of the Women Disciples*. Huntsville, AR: Ozark Mountain Publishing, 2020.

Wyatt, Nicolas. "Royal Religion in Ancient Judah." In *Religious Diversity in Ancient Israel and Judah,* edited by Francesca Stavrakopoulou and John Barton, 61–81. London: T&T Clark, 2010, 2012.

Wyatt, Tim. *The Cycles of Eternity: An Overview of the Ageless Wisdom*. Leeds, UK: Leeds Theosophical Society, 2016.

For an extended reading list see, *Mary Magdalene Unveiled*, pages 565–84.

INDEX

Please note that italicized page numbers
in this index indicate illustrations.

ABOUT THE AUTHOR

Annine E. G. van der Meer is a Dutch historian of religion who holds a Ph.D. in theology from the University of Utrecht. She was the seventh and last, as well as the only female, student to write her doctoral thesis under the guidance of the late Prof. Gilles Quispel. He was an internationally known scholar of apocryphal Christian texts, famous for his translation of the so-called fifth Gospel, the Gospel of Thomas, and other texts from the Nag Hammadi library, which were rediscovered in Upper Egypt in 1945.

Annine has written several authoritative books about the hidden history of the sacred feminine and of women and their forgotten contribution to evolution and civilization. This involves digging Her-Story out from under His-Story in order to write Our-Story. Where necessary, she integrates the established images of woman and man for the purpose of achieving equality, harmony, balance, and peace in the world. In 2008, she founded the Pansophia Foundation, a school of wisdom in the twenty-first century, which she led until 2018. Its mission was to combine the raising of consciousness with spirituality and empowerment of women. In 2015, she founded her own Dutch publishing house, Pansophia Press, in which she continues this wisdom work, supported by a large Pansophia community in The Netherlands.

Annine is the author of nineteen books in Dutch, English, French, and German. In Dutch, French, and English, she published *The Black Madonna from Primal to Final Times* (with 289 illustrations and descriptions of 450 Black Madonnas in France), published respectively in 2015, 2018, and 2020. In English in 2013 and in German in 2020 she published *The Language of MA the Primal Mother: The Evolution of the Female Image in 40,000 Years of Global Venus Art.* In 2021, she published a very successful Dutch book *Maria Magdalena Ontsluierd (Mary Magdalene Unveiled)*, a commentary about *The Gospel of the Beloved Companion* that was written by Mary Magdalene herself. In 2023, she published the English translation and updated book *Mary Magdalene Unveiled: Hidden Sources Restore Her Broken Image.*

In July 2010 the Manifest Female Energy identified and honored Annine among thirty-three world-women who had been inspired by female energy and who contributed to transformation processes in the world—a new world in which feminine and masculine energies mutually inspire each other and are growing toward a new, powerful, and creative world order.

For more information about Annine's work:

www.anninevandermeer.nl and **www.pansophia-press.nl**